DISCOVERING CALCULUS WITH THE TI-81 AND THE TI-85

DISCOVERING CALCULUS WITH THE TI-81 AND THE TI-85

Robert T. Smith
*Millersville University
of Pennsylvania*

Roland B. Minton
Roanoke College

McGraw-Hill, Inc.
New York   St. Louis   San Francisco   Auckland   Bogotá   Caracas
Lisbon   London   Madrid   Mexico   Milan   Montreal   New Delhi   Paris
San Juan   Singapore   Sydney   Tokyo   Toronto

DISCOVERING CALCULUS WITH THE TI–81 AND THE TI–85

Copyright © 1993 by McGraw-Hill, Inc. All rights reserved.
Printed in the United States of America. Except as
permitted under the United States Copyright Act of 1976,
no part of this publication may be reproduced or
distributed in any form or by any means, or stored
in a data base or retrieval system, without the prior
written permission of the publisher.

1 2 3 4 5 6 7 8 9 0 DOC DOC 9 0 9 8 7 6 5 4 3

ISBN 0-07-059199-7

The editor was Maggie Lanzillo;
the production supervisor was Friederich W. Schulte.
R. R. Donnelley & Sons Company was printer and binder.

Library of Congress Cataloging-in-Publication Data

Smith, Robert T. (Robert Thomas), (date).
Discovering calculus with the TI-81 and the TI-85 / Robert T.
Smith, Roland B. Minton.
    p.   cm.
Includes bibliographical references and index.
ISBN 0-07-059199-7
1. Calculus—Data processing.   2. TI-81 (Calculator)   3. TI-85
(Calculator)   I. Minton, Roland B., (date).   II. Title.
QA303.5.D37S56   1993
515'.0285'41—dc20   92-40469

Contents

Preface

Calculus and Calculators

Over the last ten years, the microcomputer revolution has significantly impacted virtually every aspect of business and industry. One of the more recent and significant developments in this revolution has been the graphical representation of information. Humans are accomplished visual thinkers, with highly developed abilities for pattern recognition and visual comparisons. Computers have allowed worldwide technical industries to take better advantage of these visual skills.

Today, cardiologists routinely examine multidimensional graphical displays of heart activity to "see" heart damage. Engineers now use computer generated animated graphs to study the aerodynamic characteristics of new automobile designs. Even athletic coaches are getting into the act, using animated displays of an athlete's form to improve the athlete's technique. We could go on and on, but the point is that all of these things were unavailable only a few short years ago. In today's world, an understanding of graphical information is vital. We would argue that this understanding comes from mathematics in general and calculus in particular.

The point of learning the calculus is not simply to generate correct answers to problems. Indeed, the purpose is to learn how to *think* mathematically. This book is all about *understanding* calculus, not simply as preparation for an increasingly technical job market, but also to foster a deeper appreciation of what mathematics is. In this book, we show how the graphical and numerical capabilities of the TI-81 and TI-85 advanced calculators can help students apply their innate visual skills to better understand calculus and the modern mathematical world in which they live.

As we complete this book, the TI-81 is barely three years old. Within the last several months, it has been joined by its sibling, the TI-85. The TI-85 is incredibly powerful, with a tremendous amount of built-in memory and a sophisticated programming language. Although the TI-81 has far less memory and a less sophisticated programming language, it is entirely sufficient for the tasks undertaken in elementary calculus. Both of these machines are exceptionally easy to use. We might also observe that with the ability to perform graphical as well as numerical manipulations, these machines could more correctly be referred to as hand-held computers. Used properly, the TI-81/85 can be a valuable tool in exploring the concepts of calculus.

What you find in this book is the result of our experiences in the classroom as well as in computing in general. Our discussion does not focus on the TI-81/85. Rather, we concentrate our efforts on the main concepts of the calculus, using the calculator as a tool. We have therefore emphasized only those features of the TI-81/85 that we feel aid the study of calculus. There are many other features of these machines (especially the TI-85) which we give only passing mention to or ignore altogether. For those interested in a more thorough discussion of the calculators' features, we suggest the excellent owner's manuals which come with the calculators.

Using This Book

As its size alone should indicate, this book is certainly not a complete calculus text. The reader should have a standard calculus text as well, for although much of our discussion is self-contained, we make frequent reference to material found in such a text. Our choice of topics reflects those most often found in the first two semesters of calculus which we feel most benefit from the introduction of the TI-81/85. We also provide more realistic problems, as well as the means of solving them.

You can use this book in a variety of ways. You will find, as our own students have, that you can sit down with this book and your calculator and learn calculus with the fresh approach of a new technology. Work through the book to gain that new perspective, use our exercises to supplement those of your standard calculus text, or – and this is our preferred choice – do both. No previous experience with graphing calculators is needed to use this text. We start from scratch in Chapter 1. You need only have the desire to learn some new mathematics using a slightly different approach.

We strongly recommend that you work carefully through Chapter 1 before going on to the later chapters. The material on graphing found there is prerequisite for almost everything that follows. This chapter also provides an excellent review of precalculus from a graphical perspective. Most students should develop important graphical intuition while learning to use their calculators effectively.

The TI-81 and the TI-85 are in some ways very similar machines and we most often discuss their use making no distinction between the two. In those instances where there are differences, we have tried to explain them carefully and when there is any chance for confusion, we have discussed the two machines separately. One of the most noticeable differences is that the TI-85 has a built-in equation solver. This important feature is discussed in section 1.3. In Chapter 4, we provide programs for both machines that will allow students to quickly solve equations numerically. We should also note that the TI-85 generally carries several more digits of accuracy than the TI-81. In most instances, we have displayed the TI-81 results, while the TI-85 results may be slightly more accurate. When there are significant differences, we have noted them. Other differences worth mentioning are the TI-85's substantially larger memory, its improved menu structure and its more sophisticated programming language. We illustrate other differences throughout the text. If you have not yet purchased a graphing calculator, we strongly recommend the TI-85. Its advanced features are worth the small increase in cost.

We have tried to write programs that are as easy to follow as possible. By doing so, we sometimes sacrifice a certain level of computational efficiency or sophistication, but we hope that we have better enabled you to understand the mathematics. We have included many examples throughout the text, as well as a wide variety of exercises. Of particular note are the "Exploratory Exercises" found at the end of every section. These provide good experience in tackling some extended, open-ended problems. We have also provided numerous tips on using the TI-81/85 more efficiently.

For your convenience, when we refer to a special key (or a "softkey" located in one of the many menus), we put that key in a box. For instance, $\boxed{\text{ENTER}}$ indicates that you should press the ENTER key, rather than type the letters E-N-T-E-R. The TI-85, in particular, makes extensive use of a menu structure for storing its numerous commands. In order to help you find the commands in the numerous menus, we have included a listing inside the front cover of the book (for the TI-81) and inside the back cover (for the TI-85).

Acknowledgements

We would like to express our deep appreciation to the many individuals whose assistance with this project was invaluable. Of particular note, we wish to thank the reviewers: Benjamin F. Esham (State University of New York College at Geneseo), Robert T. Moore (University of Washington) and Evelyn J. Weinstock (Glassboro State College), whose comments were extremely helpful as we revised the manuscript. We also want to thank the reviewers for our first book, *Discovering Calculus with the HP-28 and the HP-48*, whose advice continued to be of help with the preparation of the current work: Daniel S. Drucker (Wayne State University), Bruce H. Edwards (University of Florida), Michael A. Iannone (Trenton State University), Lee W. Johnson (Virginia Polytechnic Institute and State University), Harold W. Mick (Virginia Polytechnic Institute and State University), Robert T. Moore (University of Washington) and Matthew P. Richey (St. Olaf College). Mike Keenan (Roanoke College) corrected numerous mistakes in the first book.

We would especially like to thank Pegi Proffitt and Dave Stone of Texas Instruments for their exceptional support during the preparation of this book, keeping us informed about changes during the development of the TI-85 and providing us with preliminary versions of software and hardware. Their technical assistance speeded our work by months. We want to thank our original editor, Richard Wallis, and particularly our current editor, Margaret Lanzillo, who picked up the ball and kept everything going while doing two jobs. (Thanks, Maggie!) Our thanks also go to the McGraw-Hill production staff, whose efficiency quickly moved our project along. Finally, this project would not have been possible without the patience and understanding of our spouses, Pam and Jan.

The entire text has been typeset in TeX. For sharing her TeXpertise and helping to work out numerous technical problems, we want to thank Jan Minton. The TI-81 calculator graphics were produced with Texas Instruments' PC-81, the TI-85 graphics with a prototype of LINK-85 and all were printed on a Hewlett-Packard Laserjet IIIP printer.

Robert T. Smith   Roland B. Minton
Lancaster, Pennsylvania   Salem, Virginia

About the Authors

Robert T. Smith is Professor of mathematics at Millersville University of Pennsylvania, where he has taught since 1987. Prior to that, he was on the faculty at Virginia Polytechnic Institute and State University. He earned his Ph.D. in mathematics from the University of Delaware in 1982.

Dr. Smith's research interests are in the application of mathematics to problems in engineering and the physical sciences. He has published a number of research articles on the applications of partial differential equations as well as on computational problems in tomography. He is a member of the American Mathematical Society, the Mathematical Association of America and the Society for Industrial and Applied Mathematics.

Professor Smith lives in Lancaster, Pennsylvania, with his wife Pam and daughter Katie. He is an avid volleyball and an occasional tennis and softball player. Having been born and raised a fan of the Philadelphia Phillies, he knows what it is to struggle.

Roland B. Minton is Associate Professor of mathematics at Roanoke College, where he has taught since 1986. Prior to that, he was on the faculty at Virginia Polytechnic Institute and State University. He earned his Ph.D. in mathematics from Clemson University in 1982.

Dr. Minton's research interests are in stochastic systems theory and sports science. His publications include journal articles and co-authorship of a technical monograph in control theory and parameter estimation, as well as articles on the teaching of mathematics. He is a member of the American Mathematical Society, the Mathematical Association of America and the Consortium for Mathematics and Its Applications.

Professor Minton lives in Salem, Virginia, with his wife Jan, daughter Kelly and son Greg. He enjoys playing golf and tennis and collecting Greg Minton baseball cards. A love of rock and roll and the Marx Brothers helps him understand his parental responsibilities.

Professors Smith and Minton have also written *Discovering Calculus with the HP-28 and the HP-48*, as well as the forthcoming *Discovering Calculus with the Casio fx-7700 and fx-8700*, both published by McGraw-Hill.

CHAPTER
1

Overview of the TI-81 and TI-85

1.1 Introduction

Calculus is a very broad and tremendously deep study with many and varied applications. As the name implies, it is filled with calculation. You will compute velocities, areas, volumes, etc. The skills that you learn in calculus are basic tools for studying (and yes, practicing) engineering, physics, chemistry, economics and many other diverse fields. But, calculus is more than just necessary background work for the sciences. It is a fascinating subject in its own right, where geometry and algebra come together into a powerful problem solver. The mathematical theory developed here will allow you to make connections between seemingly unrelated real world problems, ultimately leading you to a deeper understanding of the world in which we live.

These are some pretty strong statements that we've made. Indeed, what we have described above is the ideal. Unfortunately, the ideal and the reality are not always quite the same. The often messy details of algebra and computation can sometimes obscure the calculus that's behind them. We cannot, nor would we want to, get rid of all the algebra and computation involved in studying calculus. Indeed, these things are necessary. However, we also don't want you to get so lost in the details that you miss the larger picture.

By giving you easy access to fast calculations and graphics, your TI-81/85 can help you to discover important relationships. Because of its speed and extremely easy interface, you can ask those "What if I did this..." questions and get an answer without a large investment of time. In short, you can *experiment*. You can focus on the questions of calculus and leave many of the details to your TI-81/85 assistant.

As is the case throughout the book, most of what follows is applicable to users of both the TI-81 and the newer TI-85 graphing calculators. These two machines share many of the same capabilities, although the keyboards are set up somewhat differently. Most significantly, the TI-85 has much more memory, many more built-in functions and programs and a more sophisticated programming language than its TI-81 sibling. In the instances where there are differences, we give the TI-81 keystrokes followed by any modifications needed for the TI-85. For clarity, we have elected to give TI-85 programs which mostly mimic TI-81 programs. Generally, we will present a TI-81 program with a step-by-step explanation of the programming commands followed by the corresponding TI-85 program. It is important to note that the TI-85 typically obtains numerical results with one or two additional digits displayed. We will usually display the results of TI-81 calculations and point out when the results from the two calculators differ significantly.

ARITHMETIC ON THE TI-81/85

The Equation Operating System which the TI-81/85 uses to perform complex arithmetic is a marvel of simplicity. You were most likely able to start computing without even looking at the manual. Essentially, you enter an expression into the calculator exactly as you would write it down on paper. You may also use any number of parentheses in an expression.

For instance, to compute $3\cos(7/3) - 5\ln(3)$, you would enter

$$3 \boxed{\text{COS}} \ (7 \ / \ 3) - 5 \ \boxed{\text{LN}} \ (3) \ \boxed{\text{ENTER}}$$

The result is -7.565335863. If you got -2.495548808, your calculator is set to degrees mode. To change to radians mode [note that we are to compute $\cos(7/3)$, not $\cos(7/3^\circ)$], press $\boxed{\text{MODE}}$, the $\boxed{\bigtriangledown}$ key twice and $\boxed{\text{ENTER}}$ to highlight radians mode (labelled RAD on the TI-81 and RADIAN on the TI-85), and finally CLEAR to return to the home screen.

Notice that the multiplications of 3 with the cosine term and 5 with the ln term are *implied*; that is, you did not need to press the multiplication key in order to indicate them. The TI-81/85 also differs from most other scientific calculators in at least two other regards. First, even after the calculation is completed, the entire expression is displayed on the screen. This feature allows for easy checking of your work. Second, when using the built-in functions sine, cosine, etc., you enter the argument *after* the function. On most scientific calculators, pressing the SIN key will compute the sine of whatever number is presently on the screen.

Next, compute $4(-2 + 3/\sqrt{11})$. Note that the square root symbol on the TI-81/85 is located above the x^2 key, so that you need to press the 2nd key followed by the x^2 key to get $\sqrt{}$. Also, to produce the minus sign in front of the 2 you must use the negation key (−) (if you accidentally press the subtraction key, you will get an error). You then enter the expression exactly as written:

$$4(-2 + 3/\sqrt{}11) \boxed{\text{ENTER}}$$

The result is -4.381863865.

NOTES:

(1) Throughout the text we will denote multiplication by * and division by /. You should note that although the keys are marked by "×" and "÷," respectively, the operations are displayed as "*" and "/" on the screen. Further, we will denote exponentiation by ∧ (e.g., x^3 will correspond to "x ∧ 3"). This usage is consistent with both the displayed character and the labeling of the keys.

(2) You do not have to enter the closing parenthesis) in the above expression. Parentheses will automatically be closed in any expression, although we will display them for completeness.

(3) The meaning of "−" must be determined from context, since there are separate keys for negation and subtraction. We will denote the keystrokes for 8^{-2} by "8∧ − 2" and the keystrokes for $8^2 - 2$ by "8∧2−2." Note that in the first expression "−" refers to the negation key (to get a negative 2) while in the second expression "−" refers to the subtraction key (to subtract 2).

While the implied multiplication is a helpful feature which will save numerous keystrokes in entering expressions, there are some pitfalls to be avoided. For
example, to compute $\frac{1}{7}\sin 5$, you might be tempted to enter

$$1/7\ \boxed{\text{SIN}}\ 5\ \boxed{\text{ENTER}}$$

This expression, however, evaluates the same as $1/(7*\sin(5))$, because the implied
multiplication $7*\sin(5)$ is performed first and then the result is divided into 1. Be
aware of such ambiguous expressions, and make careful use of parentheses to avoid
them [in this case, we want $(1/7)\sin(5)$]. At this time, discover for yourself the
difference between $\sqrt{}\ 4+5$ and $\sqrt{}(4+5)$.

There are two other features of the TI-81/85 which help to make lengthy calculations easier. After each calculation is completed, the result of that calculation is
automatically stored in a variable named Ans. The contents of this variable can be
accessed in several ways. First, if you would like to, say, divide the last computed
value by 3, simply press $/\ 3\ \boxed{\text{ENTER}}$. The screen will read Ans/3, with the result
shown on the next line of the display. You can also recall the value stored in Ans
at any time by pressing the $\boxed{\text{Ans}}$ key [located above the $\boxed{(-)}$ key]. We will make
extensive use of this feature in writing programs throughout the text.

Another helpful feature is the Last Entry function. If you press $\boxed{\text{ENTRY}}$ (located
above the $\boxed{\text{ENTER}}$ key), the TI-81/85 will display the last expression entered on the
home screen. You can then use the arrow keys to move through the expression and
edit it. This saves considerable time when carrying out repetitive calculations or
when a typing error in a lengthy calculation goes undetected. Note that on the
TI-81, the $\boxed{\triangle}$ key has the same effect as pressing $\boxed{\text{ENTRY}}$ and requires only one
keystroke.

The following examples will introduce you to some of the most commonly used
features of the TI-81/85, while exploring some interesting situations.

Example 1. Last Entry

Consider the case of the trapped video game robot. A robot starts at one end
of a corridor ($x=0$) and moves toward the other end ($x=1$; see Figure 1.1). At
the same time, the walls at the ends of the corridor begin closing in, moving at half
the robot's speed.

The robot runs into a wall at $x=2/3$ (why?), then immediately turns and goes
the other way. Since the walls are now separated by a distance of $1/3$, the next

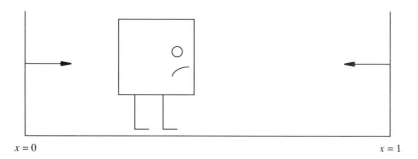

$x = 0$ $x = 1$

FIGURE 1.1

collision occurs after the robot has gone $(2/3)(1/3)$ back to the left. The second collision, then, is at $x = 2/3 - 2/9$. The walls are now separated by a distance of $1/9$. (Why?) The third collision will be at $x = 2/3 - 2/9 + 2/27$, and so on. Where exactly will the robot be trapped?

We have analyzed the problem enough to see the pattern of where the collisions occur. Let's use the TI-81/85 to crunch some numbers. First, compute $2/3$ (press 2 / 3 [ENTER]). Then subtract $2/9$ (press $-$ 2 / 9 [ENTER]). Then add $2/27$ (press $+$ 2 / 27 [ENTER]), and continue in this way to generate the following table. Note that at the start of each new calculation, your calculator inserts the variable Ans, which contains the result of the previous calculation.

Key Sequence	Result
2 / 3 [ENTER]	.6666666667
$-$ 2 / 9 [ENTER]	.4444444444
$+$ 2 / 27 [ENTER]	.5185185185
$-$ 2 / 81 [ENTER]	.4938271605
$+$ 2 / 243 [ENTER]	.5020576132
$-$ 2 / 729 [ENTER]	.4993141289
$+$ 2 / 2187 [ENTER]	.5002286237
$-$ 2 / 6561 [ENTER]	.4999237921

It looks like we're *homing in* on .5. We'll examine this idea of "homing in" (called a *limit*) very carefully in Chapter 2. We'll look further into limits involving sums in Chapter 6. For now, let's take .5 as an educated guess and rethink the problem somewhat. Is .5 a reasonable solution? The answer is "yes." Since the two walls are traveling at the same speed, they will meet in the middle at $x = .5$, with the robot trapped between them. ∎

You should notice several things about the preceding problem before going on to Example 2. We worked this problem in three stages: we performed a basic analysis, tried some calculations to estimate the solution and, finally, made an evaluation of the estimate. You should follow this process whenever possible. A numerical answer is of little value without some understanding of its meaning.

Example 2. Using User-Defined Functions

In Example 1, we were able to guess what turned out to be the precise solution by computing several points of collision and then recognizing the pattern. We will often search for an answer by repeatedly computing values of a function. For instance, we know from algebra that for

$$f(x) = \frac{1}{x^2 + 120x - 22}$$

$f(x)$ gets steadily smaller as x gets larger (for $x > 1$). Let's find a "smallness threshold," e.g., find the smallest positive integer n such that $f(n) < .001$. We'll start by computing $f(2)$ and then we'll see how the TI-81/85 can make our task easier. The keystroke sequence

$$1/(2 \wedge 2 + 120 * 2 - 22) \boxed{\text{ENTER}}$$

gives us $f(2) = .0045045045$, which is not small enough.

It looks like it will take a lot of typing to generate all of the function values we need. Fortunately, the TI-81/85 will allow us to store the expression for the function f for repeated use. The simplest way to do this is to put the expression on the "Y=" list. This is the same way that you store a function to be graphed. (We'll discuss this later.) Press $\boxed{\text{Y=}}$ on the TI-81 or $\boxed{\text{GRAPH}}$ $\boxed{\text{y(x)=}}$ (F1) on the TI-85. Then, simply enter the expression next to Y1 (or next to any other function name desired). When you are through, press $\boxed{\text{QUIT}}$ to exit the function editing screen.

NOTES:

(1) On the TI-81, be careful to press the $\boxed{\text{2nd}}$ key to execute $\boxed{\text{QUIT}}$ (since pressing $\boxed{\text{CLEAR}}$ will erase your function).

(2) For the TI-85, we will write $\boxed{\text{GRAPH}}$ $\boxed{\text{y(x)=}}$ instead of $\boxed{\text{GRAPH}}$ F1, since we want to emphasize the commands being used. A list of commands and their locations within menus is given on the back inside flap.

For the present case, enter

$$1 \, / \, (\; x \wedge 2 + 120\, x - 22\;)\;\; \boxed{\texttt{QUIT}}$$

Note that the parentheses are *not* optional here. (Why not? Which function would you end up with if you left the parentheses off?) On the TI-81, the variable X may be obtained using either the $\boxed{\texttt{X|T}}$ key or the X key (press $\boxed{\texttt{ALPHA}}$ X). Note that since the TI-85 distinguishes between upper and lower case variables, and requires the lower case variable x in functions, you *must* use the lower case x. For this purpose, it is easiest to use the $\boxed{\texttt{x-VAR}}$ key (a lower case x will be displayed on the screen).

Next, store 2 in the variable. On the TI-81, press $2\boxed{\rightarrow}$ X $\boxed{\texttt{ENTER}}$ (using the $\boxed{\texttt{X|T}}$ key for X). On the TI-85, you must enter a lower case x: press $2\boxed{\rightarrow}$ x $\boxed{\texttt{ENTER}}$ (using the $\boxed{\texttt{x-VAR}}$ key to generate a lower case x).

NOTES:

(1) Throughout the text, we will simply write $2\boxed{\rightarrow}$ x to denote these keystrokes. TI-81 users should assume that they will use the upper case X in these situations (since there is no lower case x on the TI-81).

(2) We will denote the $\boxed{\texttt{STO}\triangleright}$ key by simply $\boxed{\rightarrow}$. This is consistent with the appearance of the key on the screen and should thus help to make program keystrokes more recognizable.

Now that you have stored the function and the value of x, it remains only to evaluate the function. On the TI-81, press $\boxed{\texttt{Y-VARS}}$ (located above the $\boxed{\texttt{VARS}}$ key) and 1 (or $\boxed{\texttt{ENTER}}$) to return Y_1 to the home screen. Pressing $\boxed{\texttt{ENTER}}$ again will evaluate the expression stored in Y_1. On the TI-85, there are several ways of computing the value of the function y1 at the current value of x. The simplest is to type y1 (press $\boxed{\texttt{2nd}}$ $\boxed{\texttt{ALPHA}}$ y 1) $\boxed{\texttt{ENTER}}$ from the home screen.

The value $f(2) = .0045045045$ should be returned to the display. Try computing $f(3)$: press $3\boxed{\rightarrow}$ x $\boxed{\texttt{ENTER}}$ and evaluate Y1 again to get $f(3) = .0028818444$. Continue by computing $f(4)$, $f(5)$, and so on, until you have $f(n) < .001$. You should get $f(8) = .00099800399 < .001$. The entire sequence of calculations follows (with the results rounded off to 8 decimal places).

X-Value	Result (rounded)
x = 2	.0045045045
x = 3	.0028818444
x = 4	.0021097046
x = 5	.0016583748
x = 6	.0013623978
x = 7	.0011534025
x = 8	.00099800399

∎

We should mention at this point how to correct an expression that has been mistyped. Rather than retype the entire expression, you should simply re-enter the "Y=" list and edit the existing expression. Suppose that instead of "x∧2 + 120 x − 22" you had accidently typed

$$x \wedge 2 + 122\ x - 22$$

Press $\boxed{\text{Y=}}$ (or $\boxed{\text{GRAPH}}$ $\boxed{\text{y(x)=}}$ on the TI-85) and use the arrow keys to move the cursor over to the second 2 in 122. Pressing any key replaces the current character with the one pressed. (This is called *replace mode*: the cursor will appear as a solid flashing block.) Try this now by moving the cursor over to the second 2 in 122 and pressing 0. Press $\boxed{\text{QUIT}}$ and the corrected expression will be stored in the function Y1.

Suppose, instead, that the coefficient of x should have been 1200 instead of 120. Press $\boxed{\text{Y=}}$ and move the cursor to the spot where you want to insert the extra 0. Press the $\boxed{\text{INS}}$ key (located above the $\boxed{\text{DEL}}$ key on the TI-85); this puts the editor into *insert mode* (the cursor should now appear as a flashing underline). Pressing 0 will now insert the extra 0. Again, pressing $\boxed{\text{QUIT}}$ stores the edited expression in the function Y1.

As you edit more and more complicated expressions, you will find the need to switch back and forth between insert and replace mode while in the process of editing a single expression. This will become routine in a short time.

It may have occurred to you by now that, on such a powerful programmable calculator as the TI-81/85, there must be a better way of repeatedly computing values of a function. Of course, there is. The TI-81/85 is exceptionally easy to program. We now offer a program which will make function evaluation much simpler. We will make extensive use of this program in later chapters.

To put your TI-81 into program-edit mode, press $\boxed{\text{PRGM}}$ and then the arrow key $\boxed{>}$ (notice that EDIT is now highlighted at the top of the screen) and use the arrow keys to move the cursor down to a program number (0-9) or letter (A-Z, θ) which is not currently being used. Press $\boxed{\text{ENTER}}$ and you are in edit mode. On the first line, you may enter a name for the program (note that the calculator is already in alpha mode). We suggest that you do this routinely, using a descriptive name which will then appear in the program list (the list you see when you first press the $\boxed{\text{PRGM}}$ key).

The program listing follows, where each new line is separated by a colon (:). These are produced by pressing $\boxed{\text{ENTER}}$. When you are through entering the program, press $\boxed{\text{QUIT}}$ to return to the home screen. The $\boxed{\text{Disp}}$ and $\boxed{\text{Input}}$ commands are entered by pressing $\boxed{\text{PRGM}}$ $\boxed{>}$ (to highlight "I/O") and entering the number corresponding to the desired command. Quote marks (") are produced by pressing $\boxed{\text{ALPHA}}$ $\boxed{"}$ (above the + key). Spaces are obtained by pressing $\boxed{\text{ALPHA}}$ $\boxed{\sqcup}$ (above the 0). The function name Y_1 is found in the Y-Vars menu (press $\boxed{\text{Y-VARS}}$ 1).

:FEVAL   :$\boxed{\text{Disp}}$ "ENTER X"   :$\boxed{\text{Input}}$ X   :$\boxed{Y_1}$:$\boxed{\text{Disp}}$ $\boxed{\text{Ans}}$

Program Step	Explanation
:FEVAL	Name the program.
:$\boxed{\text{Disp}}$ "ENTER X"	Display a prompt to enter X.
:$\boxed{\text{Input}}$ X	Input the value of X.
:$\boxed{Y_1}$	Compute the value of f at the current value of x.
:$\boxed{\text{Disp}}$ $\boxed{\text{Ans}}$	Display the value of f on the home screen.

The corresponding TI-85 program is nearly identical. Press $\boxed{\text{PRGM}}$ $\boxed{\text{EDIT}}$ to enter program-edit mode and at the "NAME=" prompt, type FEVAL and press $\boxed{\text{ENTER}}$. The program follows. Be careful that you enter lower case x's and y's (pressing $\boxed{\text{2nd}}$ $\boxed{\text{ALPHA}}$ before a letter key produces a lower case letter; the $\boxed{\text{x-VAR}}$ key produces a lower case x). Note that $\boxed{\text{Disp}}$ and $\boxed{\text{Input}}$ are found in the I/O menu [press $\boxed{\text{I/O}}$ (F3)]. Quote marks are produced by pressing $\boxed{\text{I/O}}$ (F3) $\boxed{\text{MORE}}$ $\boxed{"}$ (F5). Spaces are obtained by pressing $\boxed{\text{ALPHA}}$ $\boxed{\sqcup}$ [above the (−) key]. The simplest way to enter y1 is to enter the lower case y followed by a 1.

$$: \boxed{\text{Disp}} \text{ "ENTER x"} \quad : \boxed{\text{Input}} \text{ x} \quad :y1$$

Note that the TI-85 program does not include a command analogous to the $\boxed{\text{Disp}}$ $\boxed{\text{Ans}}$ command in the TI-81 program. We have omitted this command since it would cause the TI-85 to output the word "Done" after every execution of the program, unnecessarily cluttering the display.

Both of the above programs work the same. Executing the program [on the TI-81, press $\boxed{\text{PRGM}}$, the number (or letter) of the program corresponding to FEVAL and $\boxed{\text{ENTER}}$; on the TI-85, press $\boxed{\text{PRGM}}$ $\boxed{\text{NAMES}}$ $\boxed{\text{FEVAL}}$ and $\boxed{\text{ENTER}}$] prompts the user for an x-value. Entering an x-value and pressing $\boxed{\text{ENTER}}$ returns the value of $f(x)$ to the home screen.

NOTE: After execution, FEVAL may be immediately re-run by pressing $\boxed{\text{ENTER}}$.

You should use FEVAL now to reproduce the values in the table constructed in example 2. Notice how much simpler it is to compute these values with FEVAL than it was to compute them manually.

PROGRAM NOTE: The TI-81 allows variable names consisting of only one letter (A-Z, θ). There are also only four user-defined functions (Y1-Y4) and 37 user-defined programs (labeled 0-9, A-Z, and θ). On the other hand, the TI-85 (with its considerably larger memory) allows variable names up to 8 characters in length (as long as they start with a letter and do not duplicate the name of any functions), up to 99 functions and any number of programs (identified by names of up to eight characters each) limited only by the available memory (and you will find it difficult to run out of memory).

Example 3. Parameters and FEVAL

Suppose that a ball is thrown from ground level with initial speed S ft/sec and at an initial angle x above the horizontal (see Figure 1.2).

If air effects (such as lift and drag) are ignored, an equation giving the horizontal range in feet is

$$R = \frac{S^2 \sin(2x)}{32}$$

If a ball is thrown at the initial speed of 100 mph at an angle of 30°, how far will

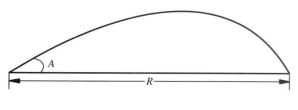

FIGURE 1.2

it go? First, make sure that your calculator is in degrees mode: press MODE and use the arrow keys to highlight "Deg" and press ENTER CLEAR . Next, enter the function into the "Y=" list as Y1. Enter the function as

$$S \wedge 2 \; \boxed{SIN} \; (2 \; x) \; / \; 32$$

and then press QUIT . Before we store a value in S, we need to convert 100 mph into ft/sec. This is done by multiplying by 5280 (feet per mile) and dividing by 3600 (seconds per hour). Press

$$100 * 5280 \; / \; 3600 \; \boxed{ENTER}$$

You should get 146.6666667. Store this in the variable S (press → S ENTER) and execute FEVAL. When prompted for an angle (x), enter 30. The range should be computed as approximately 582 ft. How much farther would the ball go if it were thrown at an angle of 40° with the same initial speed? Press ENTER to re-execute FEVAL, and enter 40 for x. We then find that the ball would go 662 ft. Notice that the difference in the horizontal distance is about 80 ft.

Now, suppose that an outfielder needs to throw a ball 300 ft, and that he/she can throw with an initial speed of 100 mph. What is the best angle of release? In this case, best would mean the smallest angle which gets the ball to its destination. (Why?) From our work above, we can conclude that 30° is too high. From the following calculations (produced using FEVAL), we conclude that the best angle would seem to be slightly greater than 13°. ∎

Angle of Release	Horizontal Distance (rounded)
$x = 20$	432.10
$x = 10$	229.91
$x = 15$	336.11
$x = 13$	294.68
$x = 14$	315.59

Note that we found an answer to this problem through a simple process of trial-and-error. In Chapter 4, we will show a more direct way of solving an equation while getting a more precise answer. For now, let's be satisfied with noting how painless the trial-and-error process was with the help of our program FEVAL.

The TI-81/85 accommodates many different levels of programming, from the crude to the very sophisticated. Since this is not intended as a text on the programming of the machine, but rather, in its use in learning calculus, we have included mostly very short and easily understood programs. We suggest some more sophisticated programs in the exercises for those who are interested in this aspect.

Example 4. Programming

In this example, we will write a *one-step* program for the problem in Example 2. An automated version of this program will be discussed in the exercises. Recalling Example 2, we would like to push one button and have the TI-81/85 calculate the value of $f(n)$ for the next value of n. We would then like to execute the program a number of times and output a sequence of values of n and $f(n)$ to the screen.

Since we want to add 1 to the value of n with each execution of the program, we use the variable Ans, which automatically stores the last number displayed on the home screen. Note that the equals sign is produced by pressing $\boxed{\text{TEST}}$ 1.

:EX114   : $\boxed{\text{Ans}}$ +1 $\boxed{\rightarrow}$ x   : $\boxed{\text{Disp}}$ "N ="
: $\boxed{\text{Disp}}$ x   : $\boxed{\text{Disp}}$ "F(N)="   : $\boxed{\text{Disp}}$ $\boxed{Y_1}$

Program Step	Explanation
:EX114	Give the program a name.
: $\boxed{\text{Ans}}$ + 1 $\boxed{\rightarrow}$ x	Add 1 to the last value (of x) and store the new value in X.
: $\boxed{\text{Disp}}$ "N =" : $\boxed{\text{Disp}}$ x	Display the new value of N (stored in the variable x).
: $\boxed{\text{Disp}}$ "F(N) =" : $\boxed{\text{Disp}}$ $\boxed{Y_1}$	Compute the value of F(N) and display it on the home screen.

The TI-85 program is similar, except that $\boxed{Y_1}$ should be replaced by y1, you must be sure that x is lower case and the display commands are combined.

:$\boxed{\text{Ans}}$ +1 $\boxed{\rightarrow}$ x   :$\boxed{\text{Disp}}$ "N=",x   :$\boxed{\text{Disp}}$ "F(N)=",y1

In order to run the program, you will need to initialize the value of Ans, by entering the starting value of n on the home screen. In this case, press 2 $\boxed{\text{ENTER}}$. You will also need to enter the function f in the "Y=" list as Y1. Executing the program will now output the value of $f(3)$ to the home screen. Pressing $\boxed{\text{ENTER}}$ will immediately rerun the program. Note that with this program, we can perform the computations needed for Example 2 in a few seconds. (Try it!)   ∎

Note that the TI-81/85 has no difficulty in referencing our user-defined function Y1, and that we wrote this program essentially by including the same keystrokes we would have used if we were performing all the steps manually. If you want to change the function (as we ask you to in the exercises) you may do so without making any changes to the program. The TI-81/85 encourages this type of structured programming: breaking up large tasks into smaller independent tasks.

PROGRAM NOTE: If your program does not work for some reason, and returns an error or simply keeps running and won't stop, you need only press the $\boxed{\text{ON}}$ key to halt execution and return to the home screen.

Exercises 1.1

In exercises 1-9, compute the indicated values. If you are familiar with exponential functions, compute the values in exercises 10-12, using the $\boxed{e^x}$ command located above the $\boxed{\text{LN}}$ key.

1. $4 + 5/12$ 2. $4/(3 - 1/6)$ 3. $7 - 4(1 + 2/3)$

4. $4^2 + 2/7$ 5. $(4 - 4/9)^2$ 6. $13.5^3 - 12^2$

7. $(3/7 + 1)^{1/2}$ 8. $(4.2 * 3.1 + 1)^{1/3}$ 9. $(3.4 * 4.1 + 1/3)^{1.2}$

10. $e^4 + 1$ 11. $(e^2 + e^{-2})/2$ 12. $e^{-3} + \ln(3)$

In exercises 13-18, compute the first 6 terms as in Example 1. If possible, guess what number the sum is "homing in" on. HINT: in exercises 16 and 18, use the factorial command ! (press $\boxed{\text{MATH}}$ 5 on the TI-81 or $\boxed{\text{MATH}}$ $\boxed{\text{PROB}}$ $\boxed{!}$ on the TI-85).

13. $\dfrac{1}{2} + \dfrac{1}{4} + \dfrac{1}{8} + \dfrac{1}{16} + \dots$

14. $1 - \dfrac{1}{3} + \dfrac{1}{9} - \dfrac{1}{27} + \dots$

15. $\dfrac{1}{1(2)} + \dfrac{1}{2(3)} + \dfrac{1}{3(4)} + \dots$

16. $\dfrac{1}{2} - \dfrac{1}{3!} + \dfrac{1}{4!} - \dfrac{1}{5!} + \dots$

17. $4 - \dfrac{4}{3} + \dfrac{4}{5} - \dfrac{4}{7} + \dots$

18. $1 + \dfrac{1}{2} + \dfrac{1}{3!} + \dfrac{1}{4!} + \dots$

19. Repeat Example 1 for $\pi/6 - \dfrac{(\pi/6)^3}{3!} + \dfrac{(\pi/6)^5}{5!} - \dfrac{(\pi/6)^7}{7!} + \dots$

In exercises 20-21, we will discover the relationship between the quadratic formula and our work in Example 2.

20. Note that $\dfrac{1}{x^2 + 120x + 22} < \dfrac{1}{1000}$ if $x^2 + 120x + 22 > 1000$ or $x^2 + 120x - 978 > 0$. Use the quadratic formula to find the two solutions of $x^2 + 120x - 978 = 0$. Use your calculator to find a decimal approximation of the larger solution. How does this relate to our answer in Example 2?

21. Use the quadratic formula as in exercise 20 to determine the smallest integer $n > 0$ such that $f(n) < 1/1000$ for $f(x) = 2/(x^2 + 10x + 100)$.

In exercises 22-24, repeat Example 2 for the given function.

22. $f(x) = \dfrac{x+1}{x^4 + 5}$

23. $f(x) = \dfrac{x-1}{x^3 + x - 2}$

24. $f(x) = e^{-x}$

In exercises 25-26, find the optimal angle, as in Example 3, for the given speed.

25. $S = 120$ mph

26. $S = 80$ mph

27. Enter the program   :EX27 :$\boxed{\text{Lbl}}$ A :EX114 :$\boxed{\text{If}}$ Y1>.001 :$\boxed{\text{Goto}}$ A and describe what it does. On the TI-81, replace :EX114 with :$\boxed{\text{PRGM}}$ # where # is the number that program EX114 is stored under. What would program EX27 do if the function f were such that Y1>.001 for all x?

28. Repeat exercises 22-24 using EX114 (Example 4) and EX27 (exercise 27). Which program do you like better?

29. When going to a bank, have you ever thought that your line moved slower than the other lines? William Feller in his classic *An Introduction to Probability Theory and Its Applications*, volume 2, devotes a section to "The Persistence of Bad Luck." Feller derives the formula $f(n) = \dfrac{1}{n(n+1)}$ for the probability that you will wait longer than the next $n - 1$ people choosing a line (under various

assumptions such as everyone receives the same service and the lines never thin out). In exercise 15, you were asked to add up 6 of these probabilities. The total probability is always 1 (100%). Did you guess 1 as the limiting value in exercise 15?

30. In the situation of exercise 29, what is the average number of people that you would wait longer than? This is computed with the *expected value* formula $f(1) + 2f(2) + 3f(3) + 4f(4) + ...$ Show that this sum keeps getting larger without ever homing in on any number. Talk about bad luck picking lines!

31. The so-called "dining room problem" is another probability problem whose solution can be found in Feller's *An Introduction to Probability Theory and Its Applications*, 3rd edition, volume 1. Suppose that n people sit down to eat and then discover that there are name tags at each seat. What is the probability that *nobody* sat down at the right place? The formula is $P = \dfrac{1}{2} - \dfrac{1}{3!} + \dfrac{1}{4!} - ... + \dfrac{1}{n!}$ (if n is even), which you were asked to compute for $n = 6$ in exercise 16. What number did you guess the sum homes in on? If the number of people for dinner becomes larger, do you think it becomes more or less likely that someone will accidentally sit in the right place? Compare your answer in exercise 16 with $e^{-1} = .367879441171....$

32. An ancient riddle is attributed to Zeno, a Greek philosopher of the fourth century B.C. A version of *Zeno's paradox* starts with Achilles 1 meter behind his rival in a race, but gaining at the rate of 1 meter/sec. Clearly, Achilles catches up in 1 sec. But, argues Zeno, before he catches up he must cut the distance to .5 meters (this takes .5 sec), and then he must cut the distance to .25 meters (this takes .25 sec), then to .125 meters, and so on. Therefore, it would seem Achilles can never catch up! The sum in exercise 13 gives the amount of time it takes to complete this seemingly endless process. Did you get 1 sec?

33. You may have already heard about the dangers of the population explosion. The following dramatic warning is adapted from the article "Doomsday: Friday 13 November A.D. 2026" by Foerster, Mora and Amiot in *Science*, volume 132, November 1960, pp. 1291-1295. Enter the function $x + .005x^{2.01}$ as Y1. We will use the program FEVAL to *iterate* this function, which projects the current population ahead one year. Start by executing FEVAL and entering 3.049 for x (the population in 1960 was about 3.049 billion). The calculator returns approximately 3.096, which is an estimate of the population in 1961. Press

ENTER to execute FEVAL again and enter Ans for x: 3.144 billion is an estimate of the 1962 population. Execute FEVAL repeatedly (entering Ans for x) and compare the equation's estimates with the actual populations shown. Then project ahead to the year 2035. Frightening, isn't it?

YEAR	POPULATION (in billions)
1970	3.721
1980	4.473
1990	5.333

34. In Example 1, we obscured a technical point of some significance. Our computations are valid only for a virtual robot of zero width. If the robot has a width of, say, .1 the first collision would occur *before* the center of the robot reaches $x = 2/3$. Repeat Example 1 using this fat robot.

EXPLORATORY EXERCISE

Introduction

In this exercise, we will take an extended look at a question that is *open-ended*; that is, there is no single correct answer or even a single correct strategy for finding a solution. The TI-81/85 makes it easy to try out ideas. This is when mathematics becomes fun! The general question to be addressed is: what is the optimal angle at which to punt a football? Mathematicians view words like "optimal" or "best" with suspicion, and a large part of this problem is to decide what should be meant by "optimal." We will assume that a ball kicked with initial speed S and initial angle A will cover a horizontal distance of $R = \dfrac{S^2 \sin(2A)}{50}$ where the 50 is intended to take into account gravity and air resistance. The ball will be in the air for $T = \dfrac{S \sin(A)}{16}$ seconds, during which time the punt coverage team will be able to run $25T + 30$ ft from the punter.

Problems

We want to compare the distances covered by the punt (R) and the punt coverage team ($25T+30$). Enter S^2 SIN (2x)/50 for Y1 and 25S SIN (x)/16+30 for Y2. Create a program F2 which is identical to FEVAL except that Y1 is replaced by Y2. We will assume that the initial speed of the punt is 90 ft/sec: enter 90 → S. Now, execute FEVAL and F2 using an angle of x=60 (degrees). We find that the punt travels about 140 ft and the punt coverage team travels about 147 ft. Since the coverage team can cover more ground than the punt, it is reasonable to assume that there would be no punt return. Now, try an angle of 45°. The punt goes 162 ft and the coverage team goes 130 ft. How far would this punt be returned? A simple rule would be to split the difference: the punt returner would make it back to the 146-ft mark. From the punt team's perspective, this is better than the 140-ft result we got from an angle of 60°. Your job is to develop a rule for deciding where the punt returner is tackled (preferably more sophisticated than the one given above) and then maximize the net distance.

Further Study

There are numerous follow-up questions for this problem. For example, we used an initial speed of 90 ft/sec in the problem above. Does the optimal angle change if you change S? Compare your results to those of others in your class, and have fun!

1.2 Graphing Capabilities

The expression "a picture is worth a thousand words" is particularly true in mathematics. For complicated problems, a graph is often useful for communicating the statement as well as the solution of a problem. Graphs also provide simple summaries of the important properties of a function. For instance, the graph displayed in Figure 1.3 (which is taken from the screen of a TI-85) shows a function which has a minimum value of approximately $y = 0$, located at about $x = 0$. We would also anticipate that the function would continue to get larger to the left and to the right of the screen displayed.

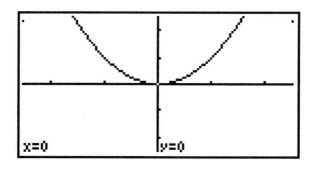

FIGURE 1.3

The TI-81/85 lets you immediately see what the graph of a function looks like. You can then use your skills with calculus to precisely label the important points on the graph, and to determine the behavior of the function in regions not displayed by the graph. This all sounds great, but there are a few significant cautions. First, your TI-81/85 does not draw graphs. All that it does is plot points, albeit lots and lots of points. You should not be so dazzled by the speed and comparative ease of use of the graphics that you lose sight of this fact. The graphs produced by the TI-81/85 are essentially a bunch of (possibly connected) dots on the screen. Even so, because of the large number of points being plotted, the graphs are quite useful and we will exploit them.

Perhaps of more significance is the fact that when you view a graph on your TI-81/85 (or any other computer generated graph), you are looking only at a small portion of the actual graph. Each graph drawn is then really a "window" through which you can see a (hopefully representative) portion of the graph. The default window shows that portion of the graph lying inside the rectangle defined by $-10 \leq x \leq 10$ and $-10 \leq y \leq 10$. Notice right away that since the vertical size of the screen is smaller than the horizontal size, this default perspective is somewhat out of proportion. Later in this section, we will discuss how to manipulate the dimensions of the graphing window.

We begin our discussion by examining the basic graphing features built into the TI-81/85. Graphs are exceptionally easy to draw and manipulate on these calculators, as we shall see. The basic graphing commands of the two machines are the same, although the names and locations within menus are somewhat different. Also, the TI-85 has some additional advanced graphing features which we will discuss later in this section and in section 1.3.

The basic graphing commands of the TI-81 are accessed from the top row of keys. The corresponding commands on the TI-85 are accessed from the top row of soft-keys (F1 thru F5) after first pressing $\boxed{\text{GRAPH}}$. The functions of the soft-keys are then displayed on the bottom line of the screen. This top row of keys has the following labels:

$$\boxed{\text{Y=}} \quad \boxed{\text{RANGE}} \quad \boxed{\text{ZOOM}} \quad \boxed{\text{TRACE}} \quad \boxed{\text{GRAPH}}$$

The only difference between the two machines is that, on the TI-85, the first soft-key is labeled $\boxed{\text{y(x)=}}$. Since there is no chance for confusion, we will henceforth denote either key by $\boxed{\text{Y=}}$ for simplicity.

Example 1. Drawing a Graph

So, what are we waiting for? Let's graph a function! Start by storing the function $f(x) = x^2 - 7$. That is, press $\boxed{\text{Y=}}$ (or $\boxed{\text{GRAPH}}$ $\boxed{\text{y(x)=}}$ on the TI-85) and enter the function (using the $\boxed{\text{X|T}}$ key on the TI-81 and the $\boxed{\text{x-VARS}}$ key on the TI-85 to produce the variable x). Enter

$$x \wedge 2 - 7$$

next to "Y1=". Now, press $\boxed{\text{GRAPH}}$. Your calculator should draw a graph like the one shown in Figure 1.4. If not, reset the graphing window to its default size (on the TI-81, press $\boxed{\text{ZOOM}}$ 6 (Standard); on the TI-85, press $\boxed{\text{ZOOM}}$ $\boxed{\text{ZSTD}}$) and redraw the graph.

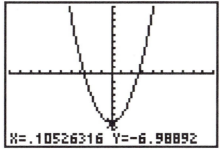

FIGURE 1.4

While the graph is displayed on the screen, the four arrow keys will move a small cross-hair (the *free cursor*) around the graphics window. As you move the cursor (try this now), the coordinates of its location are displayed along the bottom of the screen. Notice that the parabola seems to bottom out at about $y = -7$ and cross the x-axis around both $x = -2.6$ and $x = 2.6$.

One of the nicer features of TI-81/85 graphics is the trace cursor. If you press $\boxed{\text{TRACE}}$, this will place the cursor in the center of the screen and on the graph. In TRACE mode, pressing the left or right arrow keys will move the cursor *along the graph* to the left or to the right. Further, if you keep moving the trace cursor to the right (or the left) past the end of the display, the screen will scroll over to the left (or right, respectively) to follow the cursor. Note that this scrolling feature does not work with the free cursor. To erase the graph and return to the home screen, press $\boxed{\text{CLEAR}}$. (On the TI-85, $\boxed{\text{EXIT}}$ will also work and, in either case, you may have to press the key several times, to erase the menu(s) and the graph.) ∎

You should realize that the graphs produced by the TI-81/85 are not perfect representations. They quickly provide us with some idea of what the true graph looks like, but we will need to use the power of calculus to obtain precision in the areas where we need it.

Example 2. Zooming Out and RANGE

The TI-81/85 has several built-in functions used to adjust the viewing window (i.e., that small portion of the xy-plane which the calculator is currently displaying). Here, we examine the graph of $f(x) = x^2 - 7$ from several different perspectives. First, we will translate our viewing window with the RANGE command. Notice that in Figure 1.4, much of the graph is not shown due to what appear to be y-values that are off the scale (recall that the default window uses the range $-10 \le y \le 10$). We can easily adjust the range of displayed values. Press RANGE and enter -20 for Xmin, 20 for Xmax, leave Ymin $=-10$ and reset Ymax to 100 (just use the arrow keys to move the cursor to any entry that you want to change). Press GRAPH to produce the graph in Figure 1.5.

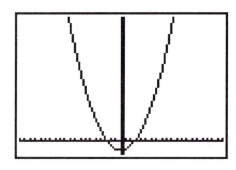

FIGURE 1.5

Notice that we are now looking at a larger piece of the graph than in Figure 1.4. The parabola has now seemingly spread out due to our having adjusted the y-range more than the x-range. In effect, we have "zoomed out" our viewing window. Using the RANGE command is the manual way to adjust the viewing window. If all we want to do is zoom out, without any interest in the particular x- and y-ranges, then we can use the ZOOM command. To discover its effect, first reset to the default viewing window by pressing ZOOM 6 (Standard) on the TI-81 or ZOOM ZSTD on the TI-85. To zoom-out, press ZOOM 3 (ZOOM Out) ENTER on the TI-81 or ZOOM

$\boxed{\text{ZOUT}}$ $\boxed{\text{ENTER}}$ on the TI-85. Notice that, in either case, the $\boxed{\text{ENTER}}$ key is pressed *after* the previous graph is re-displayed. Your display should look something like that shown in Figure 1.6.

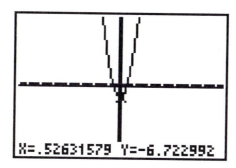

FIGURE 1.6

How far out did we actually zoom? That depends on how our zoom factors (i.e., the factors by which the x- and y-ranges were multiplied) were set. We can always press $\boxed{\text{RANGE}}$ to see what the new x- and y-ranges are. To reset the zoom factors from their default values, press $\boxed{\text{ZOOM}}$ 4 (Set Factors) on the TI-81 or $\boxed{\text{ZOOM}}$ $\boxed{\text{MORE}}$ $\boxed{\text{MORE}}$ $\boxed{\text{ZFACT}}$ on the TI-85. Notice that you may individually adjust the rate of zoom in the x- and y-directions, by setting XFact and YFact differently. For most purposes, leaving both XFact and YFact equal to some number between 2 and 4 will suffice. As you work through the examples, you should experiment using different values for these.   ∎

Example 3. ZOOM IN and ZOOM BOX

We saw two different ways to zoom out in Example 2. We will now see how to zoom in on a specific part of a graph. The most obvious way is to press $\boxed{\text{ZOOM}}$ 2 (Zoom IN) $\boxed{\text{ENTER}}$ on the TI-81 or $\boxed{\text{ZOOM}}$ $\boxed{\text{ZIN}}$ $\boxed{\text{ENTER}}$ on the TI-85. Note that this command uses the same zoom factors as does the zoom out command. With both XFact and YFact set to 2, draw the graph of $y = x^2 - 7$ using the default viewing window [press $\boxed{\text{ZOOM}}$ 6 (Standard) on the TI-81 or $\boxed{\text{ZOOM}}$ $\boxed{\text{ZSTD}}$ on the TI-85]. This should return the graph in Figure 1.4 to your screen. Now zoom in once and you should get Figure 1.7.

A more precise way to zoom in on a particular portion of a displayed graph is to use the ZOOM BOX feature. This is one of the most useful graphing routines

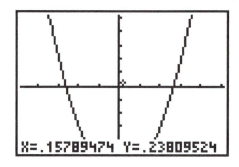

FIGURE 1.7

provided by the TI-81/85. Again, reset the the graphing window to its default specifications. Suppose that we are interested only in that portion of the graph from about −3 to 3. Pressing $\boxed{\text{ZOOM}}$ 1 (Box) on the TI-81, or $\boxed{\text{ZOOM}}$ $\boxed{\text{BOX}}$ on the TI-85, allows us to visually select a particular portion of the graph to zoom in on. Use the arrow keys to move the cursor to any corner of a rectangle enclosing the desired portion of the graph. Press $\boxed{\text{ENTER}}$. This anchors a corner of the box at the present location of the cursor. Now, move the cursor to the opposing corner of the box enclosing the desired region. As you move the cursor, you will see the box being drawn on the display. Once you have drawn an acceptable box (see Figure 1.8a), press $\boxed{\text{ENTER}}$ again, and a new graph will be drawn, one which is zoomed in on the box you drew (see Figure 1.8b).

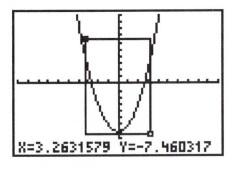

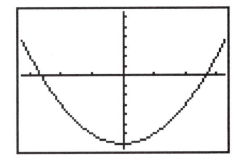

FIGURE 1.8a   FIGURE 1.8b

One thing that you should notice is that, in general, the more that we zoom in on a certain portion of a graph, the harder it becomes to see the overall (global) behavior of the function. As we will see in Example 4, perspective is an important aspect of correctly interpreting graphs.   ∎

Example 4. Asymptotes

Vertical and horizontal asymptotes are among the most recognizable features of a graph. They also provide for many students the first example of the interplay between graphs and equation solving. In this example, we will look at the graph of $f(x) = \dfrac{x-1}{x^2 - x - 2}$. You should recall that setting the numerator equal to 0 ($x - 1 = 0$) gives you the x-intercept of the graph, while setting the denominator equal to 0 ($x^2 - x - 2 = 0$) gives you the location of the vertical asymptotes, provided all common factors have already been canceled. In addition, you may have seen the idea of letting x become arbitrarily large and identifying the result with a horizontal asymptote. In this case, as x becomes very large, $f(x)$ approaches 0 and the line $y = 0$ (i.e., the x-axis) is a horizontal asymptote. Let's look at this on the TI-81/85. Press $\boxed{\text{Y=}}$ and enter

$$(\; x - 1 \;) \; / \; (\; x{\wedge}2 - x - 2 \;)$$

on the line next to Y1. Press $\boxed{\text{ZOOM}}$ 6 (Standard) on the TI-81 or $\boxed{\text{ZOOM}}$ $\boxed{\text{ZSTD}}$ on the TI-85. See Figure 1.9a for the TI-81 graph. The TI-85 graph shows slightly more detail (see Figure 1.9b). With some imagination, you should be able to visualize the vertical asymptote at $x = -1$ and possibly the vertical asymptote at $x = 2$, but the graph is just not what we expect. The problem is that we need to adjust the graphing window.

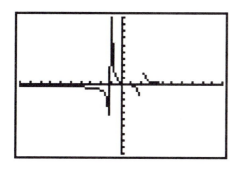

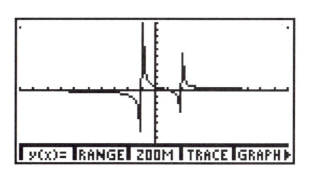

FIGURE 1.9a   FIGURE 1.9b

With a given graph, it may take some playing around with the window to see what's best. In the present case, try zooming in, but just a bit. On the TI-81, press $\boxed{\text{ZOOM}}$ 4 (Set Factors) (on the TI-85, press $\boxed{\text{ZOOM}}$ $\boxed{\text{MORE}}$ $\boxed{\text{MORE}}$ $\boxed{\text{ZFACT}}$) and then set both XFact and YFact equal to 2. Pressing $\boxed{\text{ZOOM}}$ 2 (Zoom In) $\boxed{\text{ENTER}}$ on the

TI-81 or ZOOM ZIN ENTER on the TI-85 will then produce a more recognizable graph. See Figure 1.10a for the TI-81 graph and Figure 1.10b for the TI-85 graph. This is more like what we expected from our earlier analysis.

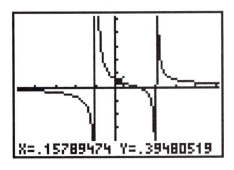

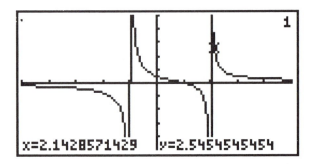

FIGURE 1.10a   FIGURE 1.10b

The graph appears to have vertical asymptotes at $x = -1$ and $x = 2$, an x-intercept at $x = 1$, and becomes horizontal to the left and to the right. So, we needed to zoom in from the initial graph to see all the features we had anticipated from our analysis. What would happen if we zoomed in further yet? Try pressing ZOOM 2 (Zoom In) ENTER on the TI-81 or ZOOM ZIN ENTER on the TI-85. (See Figures 1.11a and 1.11b for the resulting TI-81 and TI-85 graphs, respectively.)

Notice that while the detail becomes much clearer near the vertical asymptotes, the horizontal asymptotes are all but lost. The significance of Figures 1.9-1.11 is that the appearance of a graph is highly dependent on the scales you choose. It is essential to verify any features you see on a graph with calculus, because appearances can be deceiving, very deceiving.   ■

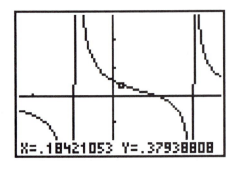

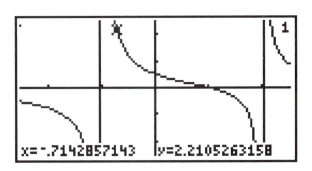

FIGURE 1.11a   FIGURE 1.11b

We note that the TI-81/85 has a scroll feature which may be convenient in exploring the horizontal asymptote. In TRACE mode, move the cursor to the far right of the screen. When you reach the edge, instead of getting "stuck," the graphing window moves eight steps over. By continuing to move the cursor to the right, you can see more and more of the graph. This feature is especially useful if you only need to see a little more to the right.

Example 5. Oblique Asymptotes

A change of scale is useful for identifying a third type of asymptote: the oblique (or slant) asymptote. For example, for $f(x) = \dfrac{2x^3}{x^2 - 1}$, by solving $x^2 - 1 = 0$ we know that there should be vertical asymptotes at $x = -1$ and $x = 1$. But what happens to $f(x)$ as x becomes large? Perhaps the graph will give us a hint. First, enter the function in the "Y=" list as Y1 (press $\boxed{\text{Y=}}$ on the TI-81 or $\boxed{\text{y(x)=}}$ on the TI-85 and enter the function on the line next to Y1)

$$Y1 = 2 \; x{\wedge}3 \; / \; (\; x{\wedge}2 - 1 \;) \;\; \boxed{\text{QUIT}}$$

Now, graph the function with the default parameters [on the TI-81, press $\boxed{\text{ZOOM}}$ 6 (standard) or on the TI-85, press $\boxed{\text{ZOOM}}$ $\boxed{\text{ZSTD}}$]. The vertical asymptotes at $x = -1$ and $x = 1$ are visible, although to see them clearly, we would need to zoom in some (the ZOOM BOX command is ideal for this). For $x < -1$ or $x > 1$ the graph seems to straighten out, but then goes off the screen (see Figure 1.12a).

To get a better idea of what goes on for x large, let's zoom out: press $\boxed{\text{ZOOM}}$ 3 (Zoom Out) $\boxed{\text{ENTER}}$ on the TI-81 or $\boxed{\text{ZOOM}}$ $\boxed{\text{ZOUT}}$ $\boxed{\text{ENTER}}$ on the TI-85. You should see the graph from about $x = -20$ to $x = 20$ (if you still have Xfact=Yfact=2). Zoom out once more and we obtain the graph in Figure 1.12b.

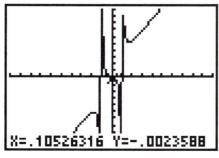

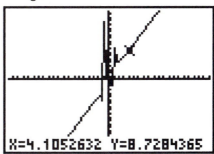

FIGURE 1.12a   FIGURE 1.12b

It is now hard to visualize the vertical asymptotes, but the graph clearly seems to straighten out to the left and to the right. Long division can help to explain what is happening. Since $\dfrac{2x^3}{x^2-1}$ can be rewritten as $2x + \dfrac{2x}{x^2-1}$, for large values of x the difference between $f(x)$ and $2x$ becomes very small. What we see in Figure 1.12b looks like the graph of $y = 2x$, for large x. To verify this, draw $y = 2x$ on top of the graph in Figure 1.12b by pressing $\boxed{\text{Y=}}$, entering the function '2x' as the function Y2 and pressing $\boxed{\text{GRAPH}}$.   ∎

It should be clear at this point that there is more to graphing than plotting points or pushing buttons on a calculator. A change of scale can dramatically alter the appearance of a graph, and the proper choice of scale depends on exactly what information is being sought.

Example 6. Intersections of Graphs

There are many instances when we are interested in finding the point(s) of intersection of two graphs. For example, one of the applications of calculus which we will see in Chapter 5 is the problem of finding the area between two curves. To do so, it will be necessary to know where the curves intersect. The TI-81/85 can draw several graphs simultaneously to help us locate any points of intersection.

For example, where do $y = x^4$ and $y = 2x + 3$ intersect? Worded differently, what are the solutions of the equation $x^4 = 2x + 3$? To get an idea of the solution, you should first graph $y = x^4$ and $y = 2x + 3$ simultaneously (simply enter these in the "Y=" list, one as Y1 and the other as Y2).

In Figure 1.13 (the TI-81 graph using the default window), we can use TRACE to locate two apparent intersections, one near $x = -1$, and a second one near $x = 1.6$.

Note that for $x = -1$, $x^4 = 2x + 3 = 1$, so that we have in fact found one intersection. We'll need to look closer to see more precisely where the second intersection is located. The simplest way of doing this is to use the ZOOM BOX command: press $\boxed{\text{ZOOM}}$ 1 (Box) on the TI-81 or $\boxed{\text{ZOOM}}$ $\boxed{\text{BOX}}$ on the TI-85. Move the cursor just below and to the left of the intersection and press $\boxed{\text{ENTER}}$. Then, move the cursor just above and to the right of the intersection and press $\boxed{\text{ENTER}}$. The box being zoomed in on is shown in Figure 1.14a. Your zoomed graph should look similar to that in Figure 1.14b.

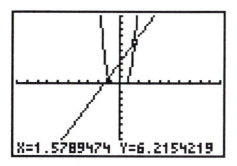

FIGURE 1.13

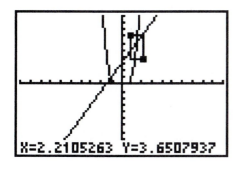

FIGURE 1.14a

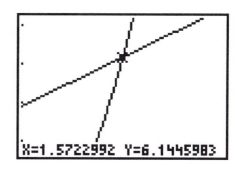

FIGURE 1.14b

Use TRACE to move the cursor to the apparent location of the intersection and read off your approximation. We got (1.5745152,6.145928); your point may differ slightly. We will discuss how to improve this approximation in section 1.3 and again in Chapter 4.   ∎

Example 7. Estimating Zeros

We could have approached Example 6 differently by looking for zeros of $x^4 - 2x - 3$, since $x^4 - 2x - 3 = 0$ is equivalent to $x^4 = 2x + 3$. We should caution that rewriting an equation can alter the type of graph that is called for. Recall that from algebra, we know that a fourth-order equation like $x^4 - 2x - 3 = 0$ can have at most 4 solutions.

Draw the graph of $y = x^4 - 2x - 3$ with the default parameters. First, store the function in the "Y=" list as Y1. Press $\boxed{\text{Y=}}$ and enter

$$x \wedge 4 - 2\,x - 3 \boxed{\text{QUIT}}$$

and then press ⃞ZOOM⃞ 6 (Standard) on the TI-81 or ⃞ZOOM⃞ ⃞ZSTD⃞ on the TI-85. As in Figure 1.14a, you can see apparent zeros near $x = -1$ and $x = 1.6$ (see Figure 1.15 for a TI-81 graph).

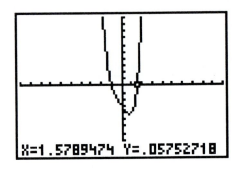

X=1.5789474 Y=.05752718

FIGURE 1.15

While we can use the TRACE command to get an idea of the location of the roots, if we want a better approximation, we should use the ZOOM BOX command to zoom in and get a closer look, much as we did in Example 6. Once you have found adequate approximations for the roots, you will need to ask whether there are more than two zeros. In the present case, we do not have any evidence indicating that there is a third or fourth zero, but we will need to use some calculus to decide for certain.   ∎

Exercises 1.2

In exercises 1-10, graph the given function and use zoom in and zoom out commands to determine how many zeros the function has. Note that in exercises 9-10, x is in radians.

1. $f(x) = x^3 + 4x^2 + x - 1$

2. $f(x) = x^3 - 5x^2 + 2x - 1$

3. $f(x) = x^3 - 3x^2 + 2x + 1$

4. $f(x) = x^4 + 3x^3 - x + 1$

5. $f(x) = x^4 - 5x^2 + 7$

6. $f(x) = x^4 + 2x^2 + 1$

7. $f(x) = x^5 - 3x^3 + 2x^2 - 1$

8. $f(x) = x^6 + 4x^5 + 2x - 1$

9. $f(x) = x^2 + 4\cos x - 1$

10. $f(x) = x^3 + x \sin x$

In exercises 11-18, graph the given combination of sines and cosines. Which graphs have simple oscillations and which have more complicated periodic patterns?

11. $f(x) = 2\sin x + \cos x$

12. $f(x) = 2\cos x - \sin x$

13. $f(x) = 3\cos 2x + \sin 2x$

14. $f(x) = 3\sin 2x - \cos 2x$

15. $f(x) = \cos x + \cos 2x$

16. $f(x) = \sin x + \sin 2x$

17. $f(x) = \cos x + \cos 3x$

18. $f(x) = \sin x + \sin 3x$

19. Based on your graphs in exercises 1-8, describe what the graph of an nth-order polynomial looks like. Make up several polynomials of your own to help see the patterns.

20. Based on your answers to exercises 1-8, how many zeros would you expect an nth-order polynomial to have? Make up several polynomials of your own to test your answer.

21. Based on your graphs in exercises 11-14, what does the graph of $A\cos wx + B\sin wx$ look like for constants A, B and w? What is the amplitude (maximum value)?

22. Based on your graphs in exercises 15-18, what does the graph of $\cos ax + \cos bx$ or $\sin ax + \sin bx$ look like for constants a and b? What is the period?

In exercises 23-28, graph the function and find all asymptotes (vertical, horizontal and oblique).

23. $f(x) = \dfrac{2x}{x^2 + 2x - 3}$

24. $f(x) = \dfrac{3x - 1}{x^2 - 4}$

25. $f(x) = \dfrac{2x^2 + 1}{x^2 + x + 1}$

26. $f(x) = \dfrac{x^3 + 4}{x^2 - 4}$

27. $f(x) = \dfrac{x^3 + 2x + 1}{x^2 + 4}$

28. $f(x) = \dfrac{x^3 - 4}{2x^3 + 7}$

In exercises 29-34, graph both functions simultaneously and use the ZOOM BOX command to estimate the points of intersection of the graphs.

29. $f(x) = \sqrt{x^3 + 2}$, $g(x) = 4x + 1$

30. $f(x) = x^4 - 3$, $g(x) = x + 2$

31. $f(x) = \cos x$, $g(x) = x$

32. $f(x) = \sin 2x$, $g(x) = x^2 - 1$

33. $f(x) = e^{-x}$, $g(x) = x$

34. $f(x) = e^x$, $g(x) = x^2 - 1$

In exercises 35-38, determine if there is an oblique asymptote. If there is not, what general shape does the graph assume? Check your answers with long division.

35. $f(x) = \dfrac{2x^3 + 2x}{x^2 + x + 1}$

36. $f(x) = \dfrac{3x^4 - 3x^2 - 1}{x^2 + 2x + 1}$

37. $f(x) = \dfrac{2x^3 - x + 1}{x + 2}$

38. $f(x) = \dfrac{3x^4 - 3x^2 - 1}{x + 2}$

39. Based on exercises 35-38, state a rule for what the general shape of $f(x) = \dfrac{p(x)}{q(x)}$ assumes if $p(x)$ and $q(x)$ are polynomials of degree n and m, respectively.

40. Graph the circle $x^2 + y^2 = 4$ by graphing $y = \sqrt{4 - x^2}$ and $y = -\sqrt{4 - x^2}$ simultaneously using the standard graphing window. Why doesn't the graph look like a circle? (**HINT**: move the free cursor around; by how much do x and y change?) Change the Range parameters to Xmin$=-4.8$, Xmax$=4.7$, Ymin$=-3.1$, Ymax$=3.2$ on the TI-81 and use the zoom command ZDECM on the TI-85. Why does the graph look like a circle now?

41. Graph the ellipse $x^2 + 4y^2 = 16$ using the graphing window of exercise 40. How does the major axis (the distance along the x-axis) compare to the minor axis (the distance along the y-axis)? Change the zoom factors to 2 for x and 1 for y, and then zoom in. What does this tell you about the relationship between a circle and an ellipse?

EXPLORATORY EXERCISE

Introduction

What is a tangent line? For many of us, the phrase "tangent line" brings to mind the image of a line resting on the outside of a circle, touching at exactly one point. Unfortunately, this image is too limited to serve as a useful working definition of the tangent line. For example, think of the curve $y = \sin x$; what is the tangent line at $x = 0$? We can't rest a line on the outside of this graph! The idea of "touches at one point" is not especially helpful, either. The line $y = -x$ touches at exactly one point (see Figure A), but it is clearly *not* what we mean by tangent line. Furthermore, it is often the case that tangent lines touch at more than one point (see Figure B for the tangent to $y = \sin x$ at $x = 1.4$). In this exercise, we will develop an accurate and useful intuition for tangent lines.

Problems

First, graph $y = x^2 - 2x + 1$ (use the Range parameters Xmin$=-4.8$, Xmax$=4.7$, Ymin$=-3.1$ and Ymax$=3.2$ on the TI-81 and use the zoom command ZDECM on the TI-85). Then, press $\boxed{\text{ZOOM}}$ $\boxed{\text{BOX}}$ and move the cursor to the point $(0,1)$. Then move the cursor 2 pixels to the left and 2 pixels down and mark this point (press $\boxed{\text{ENTER}}$). From there, move the cursor up 4 pixels and to the right 4 pixels and press $\boxed{\text{ENTER}}$ again. The graph should appear to be a straight line with $(0,1)$ at the center of the screen. Use the cursor to find a second point on this curve [we

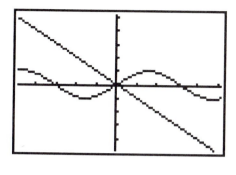

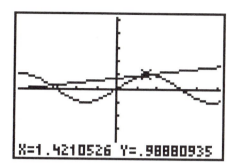

FIGURE A   FIGURE B

already know (0,1) is a point] and find the equation of the line connecting (0,1) and
the second point.

Now remove the graph and change the Range parameters back to the original
values. Draw the line found above on top of this graph (enter the equation of the
line as Y2 and graph). The line and parabola share several pixels and are graphed
with adjacent pixels in some areas. For which values of x are the two graphs close
to each other? In this region, we say that the line approximates the parabola well.
Repeat the above process with the following functions and points.

$$f(x) = x^3 \qquad \text{at } (0,0)$$
$$f(x) = \sin x \qquad \text{at } (0,0)$$
$$f(x) = \cos x \qquad \text{at } (0,1)$$
$$f(x) = \sqrt{x + 1} \qquad \text{at } (0,1)$$
$$f(x) = x^2 - 1 \qquad \text{at } (2,3)$$

The lines found above are approximations of tangent lines. Based on your work in
this assignment, describe in one or two sentences what a tangent line is.

Further Study

The tangent line problem is central to the study of calculus. You will see variations
of this problem under several names as you progress through your mathematics
courses. As presented here, the tangent line provides a *linear approximation* to
the curve. If we follow the line for a short interval and then calculate the linear
approximation at the next point, we will remain fairly close to the curve. This is
the basis for a technique called *Euler's method* used to approximate solutions of
differential equations. This is discussed in section 3.4. The slope of the tangent line
will be called the *derivative* of the function at the point. The tangent line is also an
example of a *Taylor polynomial* (to be discussed in Chapter 6).

1.3 Solving Equations

In several examples in the two preceding sections, we found approximate solutions to problems through a process of trial-and-error. The speed and the graphics of the TI-81/85 made this practical for the problems we encountered, but it is important to realize that most real world problems are more complicated and thereby more difficult to solve. A study of calculus provides us with some very effective tools for solving many such problems. Often, these methods will eventually require us to solve an equation(s). For example, we may need to find an x (a *root* or *zero*) for which $f(x) = 0$.

While a given equation may precisely define a solution to the problem at hand, we still must be able to find the solution(s) of the equation. For instance, the equation $x^3 - x^2 - 2x + 2 = 0$ has 3 solutions. One of these, $x = 1$, is easily found and easily checked (just plug in $x = 1$). Two more solutions can be found graphically. We can approximate these (try this graphically, using the TRACE command) by $x = 1.4$ and $x = -1.4$. With persistence, we can further refine our estimate of these zeros to be $x = 1.414$ and $x = -1.414$. (Do these digits look familiar?) In this case, of course, we can factor, to obtain

$$x^3 - x^2 - 2x + 2 = (x - 1)(x^2 - 2)$$

We then see that the zeros are $x = 1$ and $x = \pm\sqrt{2}$.

There are several important points to be made here. First, notice that not all answers are integers! Further, when the solutions are not integers, we often have to be content with an approximation with a fixed number of decimal places of accuracy. In many cases, there is no way to find an exact solution.

The TI-85 has several built-in features giving us some options for finding good approximations to solutions. In particular, we will discuss the use of the TI-85 SOLVER. Users of the TI-81 are encouraged to work through the problems graphically, since the solving features of the TI-85 are not available on the TI-81. We especially urge TI-81 users to work through the exercises, which involve parametric equations and systems of equations. Equation solving is important enough that we return to it in Chapter 4, where we will examine, in detail, several methods for efficiently finding accurate approximations (which work equally well on the TI-81 and the TI-85).

Example 1. The SOLVER

Let's return to the equation $x^3 - x^2 - 2x + 2 = 0$, and see how our calculators can be better used. We will start by drawing a graph. Press $\boxed{Y=}$ and enter the function on the line next to Y1:

$$x \wedge 3 - x \wedge 2 - 2x + 2 \quad \boxed{QUIT}$$

As usual, pressing $\boxed{ZOOM}$ 6 (Standard) on the TI-81 or $\boxed{ZOOM}$ $\boxed{ZSTD}$ on the TI-85 will produce the graph with the default graphing window (see Figure 1.16 for the TI-85 graph). Notice that all 3 zeros are visible. Now, using the TRACE command, move the cursor to the left until you cover up the point closest to the intersection with the x-axis.

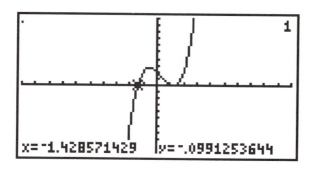

FIGURE 1.16

The root is located near $x = -1.43$. If we were to continue graphically, we would use the ZOOM BOX command to zoom in and get a more precise estimate of the root (do this now if you are using a TI-81). On the TI-85, we can instead use the built-in SOLVER: press $\boxed{SOLVER}$ (located above the $\boxed{GRAPH}$ key). On the top line of the display, enter y1 next to the label "eqn" [if there is already a function displayed on the top line, pressing $\boxed{CLEAR}$ will delete it; note that one of the soft keys (F1-F5) should correspond to y1]. This will then refer back to the function which you've already stored in the "Y=" list. Pressing $\boxed{ENTER}$ will now display the remaining variables (see Figure 1.17).

The second line of the SOLVER screen is asking for the value which you would like the expression "exp" to have. Since we're looking for roots, enter 0. On the third line, you should enter a guess as to where you think a solution may be. In this instance, enter -1.43. Finally, with the cursor still located on the x line, press

```
exp=y1
 exp=0
 x=-1.43
 bound={-1E99,1E99}



GRAPH RANGE ZOOM TRACE SOLVE
```

FIGURE 1.17

SOLVE (F5). This should return the value $x = -1.4142135623731$ to the third line. The advisory message "left−rt=0" displayed on the last line of the screen (see Figure 1.18) indicates that the TI-85 calculates y1 to be exactly equal to the value on line 2 of the SOLVER screen (i.e., 0).

```
exp=y1
 exp=0
▪x=■1.4142135623731
 bound={-1E99,1E99}
▪left-rt=0


GRAPH RANGE ZOOM TRACE SOLVE
```

FIGURE 1.18

Let's see how close the approximation is (recall that the exact roots are 1 and $\pm\sqrt{2}$). Compute $-\sqrt{2}$ now to compare with the approximate root found by the SOLVER. All of the digits displayed are correct! As we will see, the SOLVER is not always this accurate. You should always pay close attention to the advisory message to get an idea of the accuracy. Even if we did not know the precise answer, we would know that our approximation gives a very small function value.

You may wonder why we bothered drawing a graph. As we will see in our discussion in Chapter 4, it is important to start with an initial approximation close to the solution that you want. In the preceding, the point $x = -1.43$ told the calculator where to start looking for a solution. Next, replace the current value of

x with a new guess, say $x = 0$. If you press $\boxed{\text{SOLVE}}$, you should end up with the (exact) root $x = 1$. Try the initial guesses $x = 1.1$, 1.2,... until you get the third solution near $x = 1.43$.   ∎

As an alternative to the preceding method, the TI-85 has a feature which allows you to solve for roots while simultaneously looking at the graph. For the preceding problem, while the graph and graphing menu are still displayed, press

$$\boxed{\text{MORE}} \quad \boxed{\text{MATH}} \quad \boxed{\text{ROOT}}$$

This activates the TRACE cursor. Use the arrow keys to move the cursor to a point on the graph which appears to be close to a root (try the root near $x = -1.43$) and press $\boxed{\text{ENTER}}$. The calculator will display the location of an approximate root in the lower left corner of the screen (see Figure 1.19). In this case, you should see $x = -1.414213562$ and $y = 0$. The display of $y = 0$ indicates that the TI-85 thinks that it has found a root which is exact (to the number of decimal places displayed). Also, notice that this procedure gives you several fewer decimal places of accuracy than did the SOLVER.

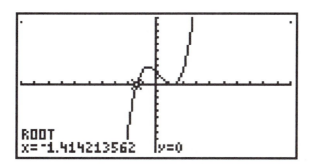

FIGURE 1.19

Example 2. Solutions of Equations

In Example 6 of section 1.2, we used graphics to look for intersections of the graphs of $y = x^4$ and $y = 2x + 3$. To find the points of intersection, we are led to solve the equation $x^4 = 2x + 3$. In section 1.2, we found two solutions graphically, one at $x = -1$ (check that this is an exact solution!) and the other near $x = 1.5$. To get greater precision than we can obtain graphically, we can use the SOLVER.

First, store the two functions in the "Y=" list, as y1 and y2, respectively. Then, press SOLVER . On the first line, enter y1−y2 and press ENTER . Then, on the next line, set exp to 0 and press ENTER . Finally, enter an initial guess for the root (here, we use $x = 1.5$) and press SOLVE . The TI-85 returns the values $x = 1.574743073887$ and left-rt = 2E−13. In this case, this is an indication that the SOLVER thinks that it has found an approximation of the intersection, where the difference between the two functions is about 2×10^{-13}.

Again, there is a graphical alternative to manually using the SOLVER. Simply graph the two functions and from the graphics window, press MORE MATH MORE ISECT . As with ROOT, this activates the TRACE cursor. Using the arrow keys, move the cursor to a point near an intersection (in this case, the one near $x = 1.5$) and press ENTER *twice*. The approximate coordinates of the intersection are then displayed across the bottom of the screen. Here, we get $x = 1.5747430739$ and $y = 6.1494861478$ (see Figure 1.20). Again, notice that there are several fewer digits displayed than are displayed by the SOLVER. Also, notice that in this instance we have no indication as to the accuracy of the answer as we did when we used the SOLVER. ∎

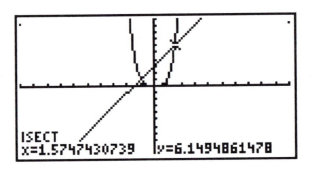

ISECT
x=1.5747430739   y=6.1494861478

FIGURE 1.20

Example 3. Finding Maxima and Minima

The extremes of functions are of special interest to us. For example, in industry, you want to maximize production while minimizing costs. As a more concrete illustration, look at the graph of $f(x) = x^3 + 4x^2 + 3x + 3$. First store the function in the "Y=" list as Y1 and press ZOOM 6 (Standard) on the TI-81 or ZOOM ZSTD on the TI-85 to draw the graph in the default window (see Figure 1.21).

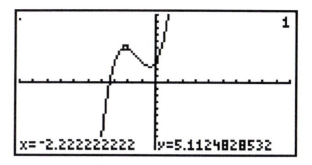

FIGURE 1.21

Notice that the graph rises to a peak, drops down to a trough, then rises again. The peak is called a *relative maximum* and the trough is called a *relative minimum*. We would like to find the coordinates of these special points. First, we will need to narrow down the interval on which the TI-85 is to look for a minimum or a maximum.

This is done by using the LOWER and UPPER commands located in the GRAPH MATH menu. The idea is to select a narrow range of x-values which include the location of the extremum (*extremum* is the generic term for maximum or minimum). From the main GRAPH menu, press $\boxed{\text{MORE}}$ $\boxed{\text{MATH}}$ $\boxed{\text{LOWER}}$, move the cursor (using the arrow keys) to a point to the left of the relative maximum and press $\boxed{\text{ENTER}}$. Next, press $\boxed{\text{UPPER}}$ and move the cursor to a point slightly to the right of the relative maximum, where there are no y-values larger than the relative maximum on the selected interval. Notice that right and left pointers are now displayed at the top of the screen indicating the interval selected with the UPPER and LOWER commands. Finally, press $\boxed{\text{MORE}}$ $\boxed{\text{FMAX}}$, move the cursor to a point near the relative maximum and press $\boxed{\text{ENTER}}$. The coordinates of the relative maximum ($x = -2.215251335$ and $y = 5.1126117909$) are now displayed across the bottom of the screen (see Figure 1.22).

This would seem to be the point that we are looking for. However, we will need some calculus to evaluate the accuracy of this answer. We will look at this question further in Chapter 4.

To locate the relative minimum, proceed as above, using the UPPER and LOWER commands to select a small interval surrounding the relative minimum and then use the FMIN command. You should find that the relative minimum occurs at about $x = -.4514162296$ and $y = 2.368869691$. ∎

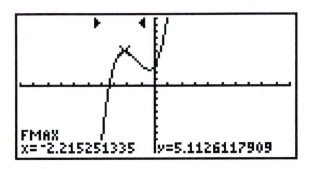

FIGURE 1.22

Example 4. Solving for One Variable in an Equation

In Example 3 of section 1.1, we used the program FEVAL to help estimate the smallest angle at which a ball could be thrown to reach a certain distance. We will now use the TI-85 SOLVER to more precisely solve that problem.

Recall that the equation $R = S^2 \sin(2x)/32$ gives the range in feet of a ball released at angle x with initial speed S ft/s. First, make sure that you switch your calculator to degrees mode (press MODE , move the cursor to "Degree" and press ENTER). Then, press SOLVER , enter the expression on the first line of the display next to the label "eqn":

$$S \wedge 2 \; \boxed{\text{SIN}} \; (\; 2 \; x \;) \; / \; 32$$

and press ENTER . Since we want to find the angle which produces a throw of 300 ft, enter 300 on the line next to "exp=." For an initial speed of 100 mph = 146.6666667 ft/s, enter 146.6666667 for S. Now, given what we found in Example 3 of section 1.1, we enter an initial guess of 13 for x and with the cursor still on the x line, press SOLVE .

The solution in degrees is given as 13.252662699755, with the advisory message "left-rt = 1E−11." This is an indication that the Solve routine could not make the expression (i.e., the horizontal range) exactly equal to 300, but with the given value of x, the difference between the range and 300 is about 1×10^{-11}. Certainly, that's close enough for a throw from center field! Further, this is much better (and faster) than the trial-and-error process described in section 1.1. The only deficiency with this process is that we can't see what the mathematical procedure is. We will discuss this further in Chapter 4.

Exercises 1.3

In exercises 1-8, find all zeros using graphics on the TI-81 or the SOLVER on the TI-85. In exercises 1-4, we indicate how many zeros there are.

1. $x^4 + x^2 - 6$ (2)

2. $x^3 + 2x^2 - 6x - 12$ (3)

3. $x^4 + 3x^3 - x + 1$ (2)

4. $x^6 + 4x^5 + 2x - 1$ (2)

5. $x^4 + x^3 - 5x - 5$

6. $4x^5 + 8x - 1$

7. $\sin x - x^2 + 1$

8. $\sin 2x - 3x + 1$

In exercises 9-10, factor the function in the given exercise to find the zeros exactly.

9. exercise 1

10. exercise 2

In exercises 11-14, find all solutions of the equations. HINT: you may want to rewrite the equation before solving.

11. $x^3 + 20 = 10x^2 + 2x$

12. $x^3 + 9x^2 = 3x + 27$

13. $\sqrt{x^2 + 4} = x^2 + 2$

14. $(x^2 - 1)^{2/3} = 2x + 1$

In exercises 15-18, find all extrema of the functions.

15. $x^3 - 3x^2 + x - 4$

16. $x^3 + 2x^2 - 2x + 1$

17. $x^4 - 7x + 2$

18. $x^4 + x^2 + 2$

In exercises 19-22, assume the showroom is to be x ft by y ft, and write down the perimeter of the required walls in terms of x and y. Since $xy = 200$ (why?) you can replace y with $200/x$ and find the desired minimum. The calculator is unusually sensitive to initial guesses in these problems, so make sure your answer is reasonable.

19. A store needs to build a showroom with 200 ft^2 of floor space. If the cost of building the showroom is proportional to the perimeter of the room (why is this reasonable?), find the dimensions of the room that minimize cost.

20. In the showroom of exercise 19, suppose that one side of the room does not need to be walled in. Find the dimensions that minimize the cost.

21. In the showroom of exercise 19, suppose that two facing walls require 3-ft openings for doors. Find the dimensions that minimize the cost.

22. In the showroom of exercise 20, suppose that two facing walls require 3-ft openings for doors. Find the dimensions that minimize the cost.

23. Use the SOLVER to find the zero of $f(x) = x^{16}$ (set $x = .5$ to start). Draw the graph of $y = x^{16}$ and try to explain why it takes so long to get the obvious answer of $x = 0$. We will discuss this type of problem in Chapter 4.

24. Find the intersections of the ellipse $\frac{x^2}{9} + y^2 = 1$ and the parabola $y = x^2 - 2$. HINT: graph the functions $Y1 = \sqrt{1 - x^2/9}$, $Y2 = -\sqrt{1 - x^2/9}$ and $Y3 = x^2 - 2$.

25. An airplane's flight path passes 100 miles east of an airport. The plane is traveling at 200 mph due south. If the airport is at (0,0) and the plane is at (100,100) at time $t = 0$, explain why the airplane's position at time t (hours) may be described by $x = 100$ and $y = 100 - 200t$. These are *parametric equations* for the plane's position. A second airplane has position $x = 50 + 150t$ and $y = 50 - 150t$. To graph the two flight paths, we use the parametric equations mode of the TI-81/85. Press MODE and move the cursor to highlight "Param." Press ENTER QUIT to return to the home screen. On the TI-81, press Y= and enter 100 for X_{1T}, $100 - 200T$ for Y_{1T} (the X|T key will now produce T's), $50 + 150T$ for X_{2T} and $50 - 150T$ for Y_{2T}. On the TI-85, press GRAPH E(t)= and enter 100 for xt1, $100 - 200t$ for yt1 (the F1 softkey will produce t's), $50 + 150t$ for xt2 and $50 - 150t$ for yt2. On both machines, we also need to adjust the Range values. Press RANGE and set Tmin=0, Tmax=1, Tstep=.01, Xmin=0, Xmax=200, Xscl=50, Ymin=-100, Ymax=100 and Yscl=50. Finally, press GRAPH to draw the paths. You will see the paths cross. At which point does this occur? Without eliminating the parameter t, ordinary algebraic methods will not help us find this point. This particular problem is easier than most because the x-coordinate of the first plane is a constant 100. At what time does the second plane reach $x = 100$? Plug this time into the y-equation of the second plane to find the point of intersection.

26. Do the planes in exercise 25 collide? To do so, their paths must pass through the same point (found in exercise 25) *at the same time.* We can get graphical evidence by setting the TI-81/85 to simultaneous graphing mode (on the TI-81, press MODE , move the cursor to Simul and press ENTER ; on the TI-85, press GRAPH MORE FORMT , move the cursor to SimulG and press ENTER). Redraw the graph. Does it look like the planes collide? To find out for sure, compare the times at which each plane passes through the point of intersection.

EXPLORATORY EXERCISE

Introduction

The equation-solving techniques discussed in this section apply to one equation for one unknown quantity. It is at least as important in applications to be able to solve *several* equations for several unknowns. The TI-81/85 is equipped with *matrix* operations to solve such systems of equations. We will explore an example of a system of equations involving the maximum speed of a car on a circular (unbanked) road. From physics, we learn that $\dfrac{F}{m} = \dfrac{v^2}{r}$ where F is the friction force, m is the mass of the car, v is the speed of the car and r is the radius of the circular path. To simplify calculations we will assume that $\dfrac{F}{m} = 100$ so that the speed of the car is given by $v = 10\sqrt{r}$. That is, speed is determined by the radius.

We will assume that we have the coordinates of 3 points on the car's path (in units of feet) which we will use to find the 3 unknowns (a, b and r) in the general equation of a circle $(x-a)^2 + (y-b)^2 = r^2$. Suppose the 3 points are (0,0), (600,600) and (520,300). To find the equation of the circle through these points, we plug them into the general equation of the circle and solve for a, b and r. With $x = y = 0$, the equation becomes $a^2 + b^2 = r^2$. With $x = y = 600$, the equation of the circle becomes $(600 - a)^2 + (600 - b)^2 = r^2$. Using $r^2 = a^2 + b^2$, we rewrite this as $600^2 - 2a600 + 600^2 - 2b600 = 0$ or, after simplifying,

$$1200a + 1200b = 720,000$$

Similarly, with $x = 520$ and $y = 300$ we get $520^2 - 2a520 + 300^2 - 2b300 = 0$ or

$$1040a + 600b = 360,400$$

This is the *system of equations* we want to solve.

Problems

We will solve these equations using a powerful technique of matrix operations. Essentially, we just copy down the numbers in the two displayed equations above into *matrix form* . We get

$$\begin{pmatrix} 1200 & 1200 \\ 1040 & 600 \end{pmatrix} \begin{pmatrix} a \\ b \end{pmatrix} = \begin{pmatrix} 720,000 \\ 360,400 \end{pmatrix}$$

It looks like we could solve for a and b by "dividing" by the matrix on the left, and what we do is analogous to division: we multiply by the multiplicative inverse. That is, if A is the matrix on the left and B is the right-hand side of the equation, the solution is $A^{-1}B$.

On the TI-81, press $\boxed{\texttt{MATRX}}$, move the cursor to "EDIT" and press 1 to edit [A]. You are prompted to enter the *size* of [A]: enter 2 (the number of rows) and 2 (the number of columns): [A] is 2×2. Now, enter 1200, 1200, 1040 and 600 (the TI-81 reads a matrix left to right, then top to bottom). Press $\boxed{\texttt{QUIT}}$ $\boxed{\texttt{MATRX}}$ and move the cursor to "EDIT." Press 2 to edit [B], enter 2 and 1 for the size of [B], enter 720000 and 360400 and press $\boxed{\texttt{QUIT}}$. Now, press $\boxed{\texttt{[A]}}$ (above the 1) $\boxed{x^{-1}}$ $\boxed{\texttt{[B]}}$ $\boxed{\texttt{ENTER}}$ and the TI-81 will return the solution.

On the TI-85, press $\boxed{\texttt{MATRX}}$ $\boxed{\texttt{EDIT}}$ and enter the name A. You are prompted for the *size* of A: enter 2 (the number of rows) and 2 (the number of columns): A is 2×2. Now, enter 1200, 1040, 1200 and 600 (the TI-85 reads a matrix top to bottom, then left to right). Press $\boxed{\texttt{EXIT}}$ $\boxed{\texttt{MATRX}}$ $\boxed{\texttt{EDIT}}$ and enter the name B. Enter 2 and 1 for the size of B, enter 720000 and 360400 and press $\boxed{\texttt{EXIT}}$. Now, press A $\boxed{x^{-1}}$ B $\boxed{\texttt{ENTER}}$ and the TI-85 will return the solution.

The first number displayed in the brackets is the value for a (about $a = .91$) and the second number is the value for b (about $b = 599$). With these values $r = \sqrt{a^2 + b^2} \approx 599$ ft and $v = 10\sqrt{r} \approx 245$ ft/sec.

How much faster could the car go if it *cut the corner?* Plot the points (0,0), (600,600) and (520,300) and compare the circle through these points to the circle through (0,0), (600,600) and (485,300). Repeat the above procedure to find the speed through the second set of points. What happens to the speed as you cut off more of the corner? Find the speed for the circle through (0,0), (600,600) and each of the following points: (400,300), (350,300) and (300,300).

Further Study
The study of matrix theory is an important part of a course in *linear algebra*. You will see pieces of linear algebra throughout calculus, particularly when you get to three-dimensional calculus.

CHAPTER

2

Numerical Computation of Limits

2.1 Conjecturing the Value of a Limit

Recall that when we say

$$\lim_{x \to a} f(x) = L$$

we mean that the function f is defined everywhere in an open interval (c, d) containing a (except possibly at a itself), and that as x gets closer and closer to a, $f(x)$ will get closer and closer to the number L (called the *limit* of f as x approaches a). We will make this intuitive notion more precise in section 2.3. For the moment, this description will do quite nicely.

You might ask why such a big deal is made about limits, anyway. After all, don't you just "plug" the value $x = a$ into the function to compute the limit? Of course, you do, for limit problems like

$$\lim_{x \to 3}(x^2 - 3x + 2) = 2$$

In fact, for any polynomial p

$$\lim_{x \to a} p(x) = p(a)$$

Unfortunately, limits are not always so easy to compute. As we will see in Example 1, we must sometimes evaluate limits which cannot be found by simply plugging into the function.

Example 1. The Limit of a Rational Function

Suppose that we next consider $\lim\limits_{x \to 3} \dfrac{x^2 - 9}{x - 3}$. We can't substitute $x = 3$ into the expression, since this would result in division by zero. Could this perhaps mean that the limit does not exist? We can use the TI-81/85 to help resolve this question by examining what happens to this function as x gets closer and closer to 3. First, we draw a graph of the function. Enter the function in the "Y=" list as Y1:

$$Y1 = (\, x \wedge 2 - 9 \,) \, / \, (\, x - 3 \,)$$

and press $\boxed{\text{ZOOM}}$ 6 (Standard) on the TI-81 or $\boxed{\text{ZOOM}}$ $\boxed{\text{ZSTD}}$ on the TI-85. The TI-85 graph is shown in Figure 2.1.

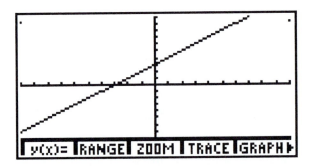

FIGURE 2.1

The graph appears to be that of a straight line about which there is nothing at all remarkable. If you press $\boxed{\text{TRACE}}$, you can use the left and right arrow keys to move the cursor along the graph while the x- and y-coordinates of the location of the cursor are displayed at the bottom of the screen (see Figure 2.2).

It should look like the function has a value around 6 when x is near 3, although there is not a point drawn at *exactly* $x = 3$. Rather than compute values of this function manually, we can use the function evaluation program FEVAL given in Chapter 1. Executing the program [on the TI-81, press $\boxed{\text{PRGM}}$, the number (or letter) of the program corresponding to FEVAL and $\boxed{\text{ENTER}}$; on the TI-85, press

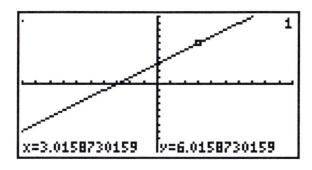

FIGURE 2.2

$\boxed{\text{PRGM}}$ $\boxed{\text{NAMES}}$ $\boxed{\text{FEVAL}}$ and $\boxed{\text{ENTER}}$] prompts the user for an x-value. Entering an x-value and pressing $\boxed{\text{ENTER}}$ returns the value of $f(x)$ to the home screen.

NOTE: After one execution, FEVAL may be immediately run again by simply pressing $\boxed{\text{ENTER}}$.

The following table is easily generated using FEVAL.

x	$f(x)$	x	$f(x)$
2.9	5.9	3.1	6.1
2.99	5.99	3.01	6.01
2.999	5.999	3.001	6.001
2.9999	5.9999	3.0001	6.0001
2.99999	5.99999	3.00001	6.00001

Notice that we have taken values of x approaching 3 both *from above* $(x > 3)$ and *from below* $(x < 3)$. (Make sure that you try entering 3 for x. What happens?) In this way, we can see if the values of $f(x)$ seem to be approaching the same number as x approaches 3 from above and from below. If not, we would say that the limit does not exist.

Both the graphical and the numerical evidence point to the conclusion that the limit is 6. In fact, for values of x closer to 3 than those shown, your TI-81/85 will report a function value of *exactly* 6. While this is indeed the correct value for the limit (we'll see why in a moment), it is not fair to say that we *conclude* that the limit is 6. We're really making more of a *guess* as to the value of the limit (although it's certainly an educated guess). The phrase that best describes what we're doing here is that we are making a *conjecture*.

You probably noticed immediately that the numerator of the fraction factors. We then have:

$$\lim_{x \to 3} \frac{x^2 - 9}{x - 3} = \lim_{x \to 3} \frac{(x - 3)(x + 3)}{x - 3} = \lim_{x \to 3}(x + 3) = 6$$

where the last limit is computed by substituting $x = 3$ (since it's the limit of a polynomial). We note that the cancellation in the preceding is valid since in the limit as x approaches 3, x gets arbitrarily close to 3, but $x \neq 3$, so that $(x - 3) \neq 0$.
∎

If you already have some experience in computing limits, you probably knew the answer to Example 1 before we even started. But, what this example should still convey to you is a process for conjecturing the value of a limit of any function which is undefined at the point in question. This is especially useful for functions whose limits are difficult to compute by hand. (It's worth noting here that a large number of functions we run into in applications fall into this category.)

A somewhat more challenging problem is the following.

Example 2. A Limit of a Product That Is Not the Product of the Limits

Consider $\lim_{x \to 0} x \sin(1/x)$. Again, we cannot resolve the limit by simply plugging in $x = 0$, since the function is not defined at $x = 0$. As with all the limit problems we'll face, we first draw a graph to get some idea of the behavior of the function near the point in question. (See Figure 2.3 for the initial TI-81 graph. *Make sure that your calculator is first set to radians mode.*)

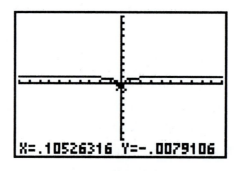

X=.10526316   Y=-.0079106

FIGURE 2.3

Although for this example the initial graph shows the behavior near $x = 0$, insufficient detail is shown to make any serious guess as to the value of the limit. Using the ZOOM BOX command, we can zoom in on the behavior near the origin (see Figures 2.4a and 2.4b for successively zoomed graphs).

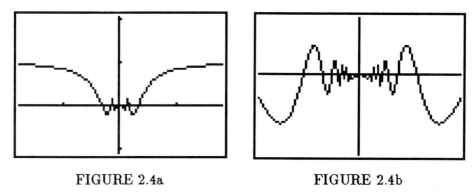

FIGURE 2.4a   FIGURE 2.4b

It appears that, although the function oscillates faster and faster as x approaches 0, the function values are getting smaller and smaller in absolute value. That is, although oscillating wildly, the function seems to be approaching 0 as $x \to 0$. Using the program FEVAL, you can quickly generate a table of values, such as the following [where the $f(x)$ values have been rounded off]:

x	$f(x)$	x	$f(x)$
.1	-5.4×10^{-2}	$-.1$	-5.4×10^{-2}
.01	-5.1×10^{-3}	$-.01$	-5.1×10^{-3}
.001	8.3×10^{-4}	$-.001$	8.3×10^{-4}
.0001	-3.1×10^{-5}	$-.0001$	-3.1×10^{-5}
1×10^{-5}	3.6×10^{-7}	-1×10^{-5}	3.6×10^{-7}
1×10^{-7}	4.2×10^{-8}	-1×10^{-7}	4.2×10^{-8}
1×10^{-9}	5.5×10^{-10}	-1×10^{-9}	5.5×10^{-10}

From both the graphical and the numerical evidence, we would conjecture that

$$\lim_{x \to 0} x \sin(1/x) = 0$$

∎

We can verify that this conjecture is true by using the following theorem, which is referred to in most calculus texts as the Pinching Theorem, the Sandwich Theorem, the Squeeze or Squeeze-Play Theorem or something to that effect.

Theorem 2.1 (Pinching Theorem) Given a function f, if we can find functions g and h such that for all x in some open interval containing a (except possibly at $x = a$), $g(x) \leq f(x) \leq h(x)$ and if $\lim_{x \to a} g(x) = \lim_{x \to a} h(x) = L$, then $\lim_{x \to a} f(x) = L$.

Figure 2.5 gives a graphical interpretation of the Pinching Theorem.

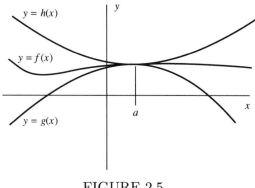

FIGURE 2.5

While this theorem is fairly easy to understand (just think in terms of Figure 2.5), to use it in practice to find limits, we must be able to dream up appropriate functions g and h. This can at times be quite a challenge. You should note, however, that for a problem where we have already made a conjecture for the value of the limit, we have a leg up on this process.

Returning to Example 2, we had conjectured that

$$\lim_{x \to 0} x \sin(1/x) = 0$$

To use the Pinching Theorem here, we need to find two functions g and h such that

$$g(x) \leq x \sin(1/x) \leq h(x)$$

for all x in some open interval containing $x = 0$, except possibly at $x = 0$, and where

$$\lim_{x \to 0} g(x) = \lim_{x \to 0} h(x) = 0$$

In order to do this, then, we will certainly have to make use of some knowledge of the sine function. One of the simplest known facts about this function is that $-1 \leq \sin(t) \leq 1$, for all t. For the case at hand we can see that $-1 \leq \sin(1/x) \leq 1$

for all x, except $x = 0$. But, we are interested in $x\sin(1/x)$. If we multiply the above inequality through by x, we get

$$-x \leq x\sin(1/x) \leq x$$

for $x > 0$, while for $x < 0$ we get

$$x \leq x\sin(1/x) \leq -x$$

Why do we get different inequalities for $x > 0$ and $x < 0$? Recall that multiplication of an inequality by a number results in the inequality reversing when that number is negative. We can summarize the above two inequalities as follows:

$$-|x| \leq x\sin(1/x) \leq |x| \quad (x \neq 0)$$

You may observe this graphically by superimposing the graphs of the functions $y = x\sin(1/x)$, $y = |x|$ and $y = -|x|$ (see Figure 2.6 for the TI-81 graph; the graphing window here is the same as for Figure 2.4b).

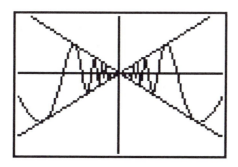

FIGURE 2.6

Clearly, $\lim\limits_{x \to 0} -|x| = \lim\limits_{x \to 0} |x| = 0$. From the Pinching Theorem, then, we also have that

$$\lim_{x \to 0} x\sin(1/x) = 0$$

as we had conjectured. That was quite a long process, wasn't it? If you haven't already done so, you should ask now whether it was worth it or not. After all, from the graphical and numerical evidence, we had a pretty good idea that the limit was 0. Why was it necessary to *prove* that the limit was indeed 0? The answer is that the more complicated the problem is, the less helpful and accurate intuition is. Computations and graphs can also be deceiving, as we'll see in section 2.2. The only way to be *certain* of the value of a given limit is to prove it.

At this point, you should recognize that there are really two somewhat different reasons for conjecturing the value of a limit. The obvious reason is that we'd like to find an approximate answer to a problem to which we cannot seem to find (at least immediately) an exact answer by hand. An equally important reason is that we would like to gain sufficient insight into a limit problem that we might discover the precise value. This second reason is a bit more difficult to get your hands on, but is a very valid reason for using a graphing calculator in exploring limit problems. We'll pursue this further in later examples and in the exercises at the end of the section.

The following is a particularly useful limit to know. It arises in the derivation of the derivative of $\sin x$.

Example 3. Limit of a Quotient That Is Not the Quotient of the Limits

Consider $\lim\limits_{x \to 0} \dfrac{\sin x}{x}$. The initial graph produced by the TI-81/85 [i.e., the graph produced by pressing $\boxed{\text{ZOOM}}$ 6 (Standard) on the TI-81 or $\boxed{\text{ZOOM}}$ $\boxed{\text{ZSTD}}$ on the TI-85] is rather flat (see Figure 2.7a for the TI-81 graph), and this makes it difficult to see what value $f(x)$ is approaching as x approaches 0. If we use the $\boxed{\text{ZOOM}}$ $\boxed{\text{BOX}}$ command to zoom in on the part of the graph of interest, we produce a graph like that in Figure 2.7b. Pressing $\boxed{\text{TRACE}}$ and moving the cursor along the graph toward $x = 0$ seems to indicate that the function approaches 1 as x approaches 0.

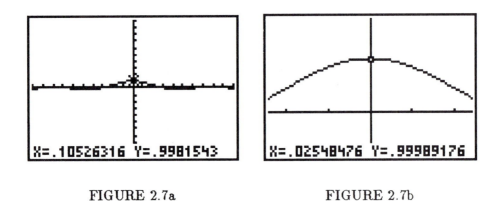

FIGURE 2.7a   FIGURE 2.7b

Using the program FEVAL, we produce the following table of TI-81 values (TI-85 users will get slightly different values due to the greater accuracy of TI-85 calculations).

x	$f(x)$	x	$f(x)$
.1	0.9983341665	−.1	0.9983341665
.01	0.9999833334	−.01	0.9999833334
.001	0.9999998333	−.001	0.9999998333
.0001	0.9999999983	−.0001	0.9999999983
.00001	1.0	−.00001	1.0

Certainly, we could compute more values, but this, together with Figure 2.7b, would seem to be sufficient evidence to warrant the testing of the conjecture:

$$\lim_{x \to 0} \frac{\sin x}{x} = 1$$

We refer the reader to the derivation of this limit found in any standard calculus text. There, as a preliminary step to finding the derivatives of the sine and cosine functions, the above limit is found, usually using a geometric argument and the Pinching Theorem. We should mention that such a proof is constructed only after one has obtained experimentally, as above, a conjectured value. Unfortunately, most standard calculus texts make no mention at all of how someone first discovered the value of the limit. Consequently, many students of calculus see such theorems as wholly unmotivated, formal exercises. We hope to help change that perception.

Example 4. A Limit with a Non-Integer Value

Consider $\lim\limits_{x \to 0} \dfrac{x}{\sin 3x}$. The initial graph drawn by the TI-81/85 is just a bit wild (see Figure 2.8a for the TI-85 graph). If you use $\boxed{\text{TRACE}}$, it should suggest that the limit is around .33. Because of the wild, oscillatory behavior, we use the $\boxed{\text{ZOOM}}$ $\boxed{\text{BOX}}$ command to zoom in on where the graph seems to cross the y-axis (Figure 2.8b). Again, using $\boxed{\text{TRACE}}$ we are led to believe that the value of the limit is around .33.

Using FEVAL, we construct the following table:

x	$f(x)$	x	$f(x)$
.1	.3383863362	−.1	.3383863362
.01	.3333833386	−.01	.3333833386
.001	.3333338333	−.001	.3333338333
.0001	.3333333383	−.0001	.3333333383

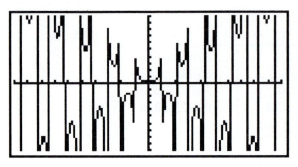

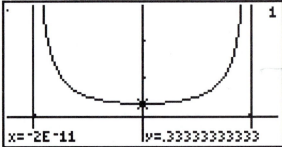

FIGURE 2.8a   FIGURE 2.8b

and so on. The graphical and numerical evidence here may seem to be less convincing than for earlier examples. One reason might be that the values are not approaching a whole number. If this were a homework problem, would this set off an internal alarm? (There must be something wrong. The answer looks too messy!) Students often learn a very subtle lesson from solving too many textbook problems whose answers are whole numbers. That is, we come to *expect* nice looking answers. When we don't get them, we start looking for a mistake. The unfortunate reality is that in real world applications of mathematics, we only very rarely run across problems which have whole number answers. Thus, we need to be practiced at solving more than just the usual unrealistic, but nice, problems.

From the preceding evidence, there are two reasonable ways in which we might solve the limit problem. First, from the evidence, we might say that the value of the limit is *approximately* equal to .3333333383. Second, in light of the expectation of a nice answer, we could leap to the conjecture that:

$$\lim_{x \to 0} \frac{x}{\sin 3x} = \frac{1}{3}$$

Of course, when it's possible to make and prove a conjecture, this is always preferable. We can prove the preceding conjecture, as follows. It should occur to the reader that the current problem is similar to Example 3. Here though, we have $\sin(3x)$ instead of $\sin(x)$. If you think that it sounds like a change of variable might be in order, you're right. Let $u = 3x$. Then $x = u/3$ and as x tends to 0, u tends to 0, also. We get

$$\lim_{x \to 0} \frac{x}{\sin 3x} = \lim_{u \to 0} \frac{u/3}{\sin u} = \frac{1}{3} \lim_{u \to 0} \frac{u}{\sin u} = \frac{1}{3}$$

as conjectured, where we have used the fact, discussed in the Example 3, that

$$\lim_{x \to 0} \frac{\sin x}{x} = 1$$

∎

Example 5. A Limit with an Irrational Value

Consider $\lim_{x \to 0} \dfrac{x}{\sin \pi x}$. From the initial TI-81 graph (Figure 2.9a) and the graph obtained from zooming in (using ZOOM BOX) on the section of the y-axis in question (Figure 2.9b), we get the idea that the limit is somewhere around .32.

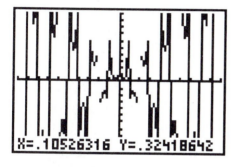
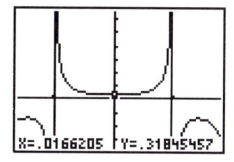

FIGURE 2.9a   FIGURE 2.9b

Using FEVAL, we generate the following table. (Note that the symbolic constant π is located above the ∧ key.)

x	$f(x)$	x	$f(x)$
.1	.3236067977	−.1	.3236067977
.01	.3183622521	−.01	.3183622521
.001	.3183104098	−.001	.3183104098
.0001	.3183098914	−.0001	.3183098914
.00001	.3183098862	−.00001	.3183098862
.000001	.3183098862	−.000001	.3183098862

and so on. As someone trained to look for whole number answers, you might be at great pains to arrive at a meaningful conjecture. For the moment, we need to be satisfied with the suggestion that the limit is approximately .3183098862. In the exercises, we shall see how to arrive at a meaningful conjecture for this problem. Looking back at Example 4 might give you an idea as to how this might be accomplished.

∎

Example 6. A Limit of a Sum Where Neither Limit Exists

Find $\lim_{x \to 2} f(x)$ where $f(x) = \dfrac{(x-1)^{1/2}}{x^2 - 4} - \dfrac{1}{x^2 - 4}$. First note that the limits of the individual terms do not exist. The elementary rule that

$$\lim_{x \to a}[f(x) + g(x)] = \lim_{x \to a} f(x) + \lim_{x \to a} g(x)$$

applies only when all three of the limits exist. This is not the case here and we must explore further to see what the limit might be.

The initial graph doesn't show any plotted points. If we use $\boxed{\text{ZOOM}}$ $\boxed{\text{BOX}}$ to zoom in on a very narrow band surrounding the x-axis (see Figures 2.10a and 2.10b for successively zoomed graphs) we see that the function appears to be approaching a value somewhere around .127.

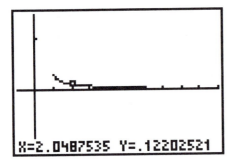

X=2.0487535 Y=.12202521

FIGURE 2.10a

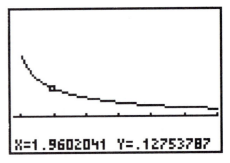

X=1.9602041 Y=.12753787

FIGURE 2.10b

Using FEVAL, we generate the following table of TI-81 values (TI-85 values are similar).

x	$f(x)$	x	$f(x)$
1.9	.1315812871	2.1	.1190459711
1.99	.1256281427	2.01	.1243781075
1.999	.1250625313	2.001	.1249375311
1.9999	.125006251	2.0001	.124993751
1.99999	.12500063	2.00001	.12499938
1.999999	.125	2.000001	.125
1.9999999	.125	2.0000001	.125

We are led to the conjecture

$$\lim_{x \to 2} \frac{(x-1)^{1/2}}{x^2 - 4} - \frac{1}{x^2 - 4} = .125$$

It is left as an exercise to show that the conjecture is correct. This requires only some elementary algebraic manipulation.   ∎

Many functions are not nearly so well behaved as those in the preceding examples. We need to be able to recognize when a function blows up at a point, as well as when it approaches a finite limit.

Example 7. A Function Whose Graph Has a Vertical Asymptote

Consider $\lim_{x \to 0} \dfrac{1}{\sin x}$. Almost any graph of the function (for instance, see Figure 2.11a and 2.11b, for the initial and zoomed-in TI-81 graphs, respectively) will seem to indicate that the function values go off the scale in the positive direction as $x \to 0$ ($x > 0$) and go off the scale in the negative direction as $x \to 0$ ($x < 0$).

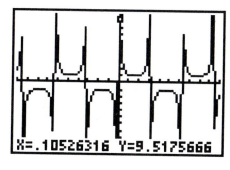

FIGURE 2.11a   FIGURE 2.11b

Using FEVAL, we generate the following table [where the $f(x)$ values have been rounded off]:

x	$f(x)$	x	$f(x)$
.1	10	−.1	−10
.01	100	−.01	−100
.001	1000	−.001	−1000
.0001	10000	−.0001	−10000
.00001	100000	−.00001	−100000
.000001	1000000	−.000001	−1000000

and so on. Since the function does not seem to be approaching any fixed value as

$x \to 0$, we are led to conjecture that the limit does not exist. To be more specific, since the function grows larger and larger in absolute value, seemingly without bound, as x gets close to 0, we conjecture that

$$\lim_{x \to 0+} \frac{1}{\sin x} = +\infty \qquad \text{and} \qquad \lim_{x \to 0-} \frac{1}{\sin x} = -\infty$$

where by this, we mean that the limit does not exist, but more specifically, it doesn't exist because the function values are growing large, in absolute value, without bound. Recall that, in this case, the graph is said to have a *vertical asymptote* at $x = 0$. Verifying these conjectures takes a bit more work than our earlier examples and we omit this.   ∎

Now that we have explored limits which tend to ∞, the seasoned student might guess (or perhaps conjecture) that we would next examine the limiting behavior of functions as x tends to infinity. We begin this with an obvious first example.

Example 8. A Function Whose Graph Has a Horizontal Asymptote

Consider $\lim\limits_{x \to \infty} \frac{1}{x}$ and $\lim\limits_{x \to -\infty} \frac{1}{x}$. The graphs produced by the TI-81/85 (Figure 2.12 shows the initial TI-81 graph) seem to indicate that the function approaches the x-axis as x gets larger and larger. (Recall that such a line is called a *horizontal asymptote*.) Figure 2.12 shows the horizontal asymptote $y = 0$.

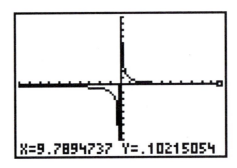

FIGURE 2.12

A table of values is easily constructed (even by hand).

x	$f(x)$	x	$f(x)$
10	.1	−10	−.1
100	.01	−100	−.01
1000	.001	−1000	−.001
10000	.0001	−10000	−.0001
100000	.00001	−100000	−.00001

It should now be intuitively quite clear that as x gets larger and larger in absolute value, $1/x$ will get closer and closer to 0. We then have that

$$\lim_{x\to\infty}\frac{1}{x}=0 \qquad \text{and} \qquad \lim_{x\to-\infty}\frac{1}{x}=0$$

It is now easy to conclude (see the exercises) that, for any positive integer k,

$$\lim_{x\to\infty}\frac{1}{x^k}=0 \qquad \text{and} \qquad \lim_{x\to-\infty}\frac{1}{x^k}=0$$

We can use these two limits to solve a large group of limit problems.   ■

Example 9. A Limit of the Form ∞/∞

Consider $\lim\limits_{x\to\infty}\dfrac{x^2+5x-7}{3x^2+4x+9}$. This is a limit of a quotient, but the limits in the numerator and the denominator are both infinite. At first glance, then, this limit has the *indeterminate form* ∞/∞ (i.e., we cannot initially determine what the limit is or even if it exists). In the initial TI-81 graph (see Figure 2.13a), the function seems to tail off almost to a horizontal line, as x approaches ∞. To get a better idea, press RANGE , set Xmin to 0, Xmax to 100, Ymin to −1 and Ymax to 1. This produces the graph in Figure 2.13b. (Note that we adjust the y-range in order to see more of the variation in the y-direction.) If we reset the x-range to Xmin = 100 and Xmax = 1000, we produce the graph in Figure 2.13c.

Using the TRACE function to follow along the graph in Figure 2.13c suggests that the function tends to a limit of about .33 as x tends to ∞. As usual, a table of values is helpful.

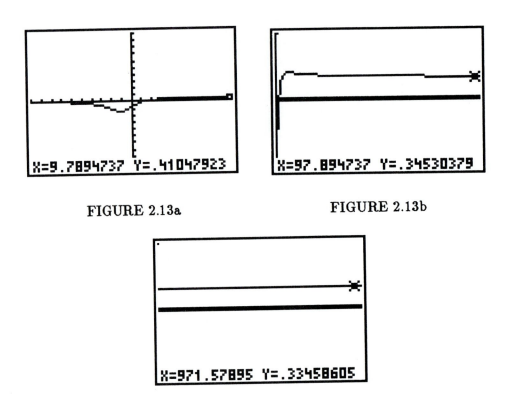

FIGURE 2.13a   FIGURE 2.13b

FIGURE 2.13c

x	$f(x)$
10	.4097421203
100	.3450623171
1000	.3345505956
10000	.3334555059
100000	.3333455551
1000000	.3333345556
10000000	.3333334556

From this evidence, we might conjecture that

$$\lim_{x \to \infty} \frac{x^2 + 5x - 7}{3x^2 + 4x + 9} = \frac{1}{3}$$

It's not hard at all to verify this conjecture. A rule of thumb for dealing with limits of rational functions (i.e., quotients of polynomials) as $x \to \infty$ or $x \to -\infty$ is to divide the numerator and denominator of the fraction by the highest power of x which appears in the denominator. For the present example, this means that we should divide top and bottom by x^2. We get

$$\lim_{x \to \infty} \frac{(x^2 + 5x - 7)/x^2}{(3x^2 + 4x + 9)/x^2} = \lim_{x \to \infty} \frac{1 + 5/x - 7/x^2}{3 + 4/x + 9/x^2} = \frac{1}{3}$$

as conjectured.   ■

We note that the TI-81/85 has a scroll feature which may be convenient in investigating limits as $x \to \infty$. In TRACE mode, move the cursor to the far right of the screen. When you reach the edge, instead of getting "stuck," the graphing window moves eight steps over. By continuing to move the cursor to the right, you can see more and more of the graph. This feature is especially useful if you only need to see a little more to the right. In investigating a limit as $x \to \infty$, using the scroll is tedious unless you first alter the x-range to a very wide view (such as Xmin=0 and Xmax=100 as above).

Example 10. A Limit Involving a Square Root

Consider $\lim\limits_{x \to -\infty} \dfrac{(x^2 - 7)^{1/2}}{2x + 1}$. Note that, like the last example, this is of the indeterminate form ∞/∞. Figure 2.14a shows the initial TI-81 graph of this function. We reset the x- and y-ranges to Xmin=-1000, Xmax=0, Ymin=-1 and Ymax = 1 (see Figure 2.14b), to try to observe the limiting behavior.

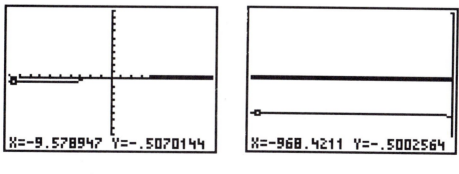

FIGURE 2.14a   FIGURE 2.14b

From the graphs (again using the TRACE function), it appears that the function has a horizontal asymptote of $y = -.5$, as x tends to $-\infty$. More convincing yet is the table:

x	$f(x)$
-10	$-.5075605664$
-100	$-.5023366526$
-1000	$-.5002483742$
-10000	$-.5000249837$
-100000	$-.5000024998$
-1000000	$-.50000025$
-10000000	$-.500000025$

It is now reasonable to conjecture that

$$\lim_{x \to -\infty} \frac{(x^2 - 7)^{1/2}}{2x + 1} = -\frac{1}{2}$$

We leave it as an exercise to show that this conjecture is true. One need only apply the rule of thumb described in Example 9, but very carefully. (HINT: Divide numerator and denominator by x and recall that $\sqrt{x^2} = |x|$.)   ∎

NOTE: We want to emphasize the interplay between the graphics, the numerical computation and the analysis of the conjecture. You might be tempted to forget about the graphs in practice. After all, just how much information have we obtained from them, anyway? It's true that the graphs which a graphing calculator generates are far too crude most times to be able to obtain even a reasonably accurate guess as to the value of a limit. So why bother with them at all? This is easy to answer. The graph gives us some intuition, an expectation of what a reasonable answer should be. If the numbers generated suggest a limit consistent with what we expect from the graphs, then we can be comfortable with our approximation or with our conjecture. However, if the limiting value suggested by a table is far out of line with our expectation, then we have serious cause for concern that one or both of the numbers or the graphs are misleading. In this case, the problem requires further analysis.

Example 11. A Limit Requiring the Use of Computation and Graphics

As an extreme example of what can go wrong with mindlessly computing a limit from a table of values alone, we offer the following. Consider $\lim_{x \to \infty} (x - \pi)\cos(\pi x)$. There is nothing particularly unusual about this function. We can easily construct the following table of values by using FEVAL (for simplicity, the $f(x)$ values have been rounded).

x	$f(x)$
10	7
100	97
1000	997
10000	9997
100000	99997
1000000	999997
10000000	9999997

From this table, we are led to the (obvious?) conjecture:

$$\lim_{x \to \infty} (x - \pi) \cos \pi x = \infty$$

On the other hand, if we use a different set of x values for our table, still tending to infinity, we get (again, the $f(x)$ values have been rounded):

x	$f(x)$
9	−6
99	−96
999	−996
9999	−9996
99999	−99996
999999	−999996
9999999	−9999996

This set of values might lead one to the seemingly reasonable conjecture that

$$\lim_{x \to \infty} (x - \pi) \cos \pi x = -\infty$$

Both of these conjectures cannot be correct! In fact, *neither* of them is correct. If we had taken the trouble to first examine the graph of the function, we might have noticed that the function exhibits a great deal of oscillation as $x \to \infty$. See Figures 2.15a-2.15d for TI-85 graphs; 2.15a is the initial graph; 2.15b was produced by resetting the x-range to Xmin = 0 and Xmax = 20; 2.15c and 2.15d are produced by repeatedly using ZOOM ZOUT (with zoom factors XFact = 4 and YFact = 4). This corresponds to ZOOM 3 (Zoom Out) on the TI-81. Having then constructed

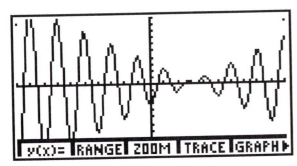

FIGURE 2.15a

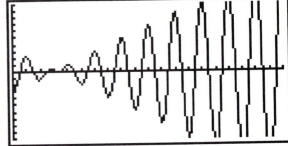

FIGURE 2.15b

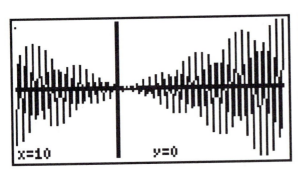

FIGURE 2.15c

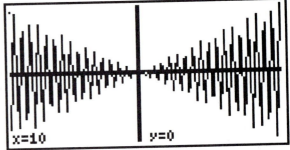

FIGURE 2.15d

either or, better yet, both of the above tables, we could correctly observe that the limit does not exist, but does not tend to $+\infty$ or to $-\infty$.

■

Exercises 2.1

In exercises 1-6, use graphics and FEVAL to conjecture the value of the limit, or conjecture that the limit does not exist.

1. $\displaystyle\lim_{x\to 1}\frac{x-1}{x^2-4x+3}$

2. $\displaystyle\lim_{x\to 0}\frac{x+1}{x^2+2x+3}$

3. $\displaystyle\lim_{x\to 0}\frac{x^2+x}{\sqrt{x^3+2x^2}}$

4. $\displaystyle\lim_{x\to 1}\frac{x^2-x}{\sqrt{x^3-x^2-x+1}}$

5. $\displaystyle\lim_{x\to 0+}\frac{\sin x}{\sqrt{x}}$

6. $\displaystyle\lim_{x\to 0}\frac{1-\cos 2x}{x}$

In exercises 7-12, conjecture the value of the limit. Then verify your conjecture by factoring or using the Pinching Theorem.

7. $\lim\limits_{x\to 2}\dfrac{x^2 + 2x - 8}{x^2 - 4}$

8. $\lim\limits_{x\to 1}\dfrac{x^3 - x^2}{x^2 + x - 2}$

9. $\lim\limits_{x\to 0} x^2 \sin(1/x)$

10. $\lim\limits_{x\to 0} x \cos(1/x)$

11. $\lim\limits_{x\to 0}\dfrac{x}{x^2 + 4}$

12. $\lim\limits_{x\to 0} x \sqrt{x^2 + 4}$

In exercises 13-16, you are asked to discover some general limit rules.

13. Conjecture the value of each limit (assume $c \neq 0$). $\lim\limits_{x\to 0}\dfrac{x}{\sin 4x}$, $\lim\limits_{x\to 0}\dfrac{x}{\sin \pi x}$, $\lim\limits_{x\to 0}\dfrac{x}{\sin cx}$. Verify your conjectures as in Example 4.

14. Conjecture the values of the limits or conjecture that they do not exist: $\lim\limits_{x\to 0} x^2$, $\lim\limits_{x\to 0} x^{1/2}$, $\lim\limits_{x\to 0} x^{-2}$. State a rule giving the various cases for evaluating $\lim\limits_{x\to 0} x^k$.

15. Conjecture the values of the limits or conjecture that they do not exist: $\lim\limits_{x\to\infty} x^2$, $\lim\limits_{x\to\infty} x^{1/2}$, $\lim\limits_{x\to\infty} x^{-2}$. State a rule giving the various cases for evaluating $\lim\limits_{x\to\infty} x^k$.

16. Conjecture the values of the limits or conjecture that they do not exist: $\lim\limits_{x\to\infty}\dfrac{e^x}{x^k}$, $\lim\limits_{x\to\infty}\dfrac{\ln x}{x^k}$ for various values of k. What does this tell you about the relative "size" of logarithms, polynomials and exponential functions?

In exercises 17-24, conjecture the value of the limit or conjecture that the limit does not exist.

17. $\lim\limits_{x\to 1}\dfrac{\sqrt{x} - 1}{x^2 + 1}$

18. $\lim\limits_{x\to 1}\dfrac{\sqrt{x + 3} - 2x}{x^2 + 2x - 3}$

19. $\lim\limits_{x\to 1}\dfrac{x^3 - 1}{\sqrt{x} - 1}$

20. $\lim\limits_{x\to 1}\dfrac{\sqrt{x} - 1}{(x - 1)^2}$

21. $\lim\limits_{x\to 0}\dfrac{\sin x^2}{x^2}$

22. $\lim\limits_{x\to 0}\dfrac{\cos x}{x}$

23. $\lim\limits_{x\to 0+} x^x$

24. $\lim\limits_{x\to 0+} (1 + x)^{1/x}$

25. At 10% annual interest compounded n times per year, an investment of $1000 is worth $\$1000\left(1 + \dfrac{.10}{n}\right)^n$ after one year. Compare the value of this invest-ment with annual compounding ($n = 1$), semi-annual compounding ($n = 2$), quarterly compounding ($n = 4$), monthly compounding ($n = 12$), daily com-pounding ($n = 365$) and continuous compounding (the limit as $n \to \infty$).

26. Repeat exercise 25 with 8% interest: the formula becomes $\$1000\left(1 + \dfrac{.08}{n}\right)^n$.

27. Argue that $\lim\limits_{x \to 0+} (1 + x)^{1/x} = \lim\limits_{n \to \infty} \left(1 + \dfrac{1}{n}\right)^n$. The second limit corresponds to one year of 100% interest compounded continuously (see exercises 25-26). Compare this limit (computed in exercise 24) to the irrational number e. What powers of e correspond to the limits $\lim\limits_{n \to \infty} (1 + .10/n)^n$ and $\lim\limits_{n \to \infty} (1 + .08/n)^n$ in exercises 25-26?

28. Verify that the limit in Example 6 equals $\frac{1}{8} = .125$. Start by finding a common denominator, then multiply numerator and denominator by $(x - 1)^{1/2} + 1$ and simplify. After cancelling out factors of $x - 2$, simply plug in $x = 2$.

EXPLORATORY EXERCISE

Introduction

What is the top speed of a human being? It has been estimated that Carl Lewis reached a peak speed of 28 mph while winning a gold medal at the 1992 Olympics (in fact, Lewis averaged 25.4 mph for the 100 meters he ran in the relay race). As noted below, to compute an average speed we need to measure distance and time. Suppose we videotape an athlete running on a well-marked track. Many video cameras record at 30 frames per second, so by counting frames we can measure time. The marks on the track help us see the distance. Combining time and distance, we get average speed!

Problems

Suppose that we collect the following data for a runner. Using the formula

$$\text{Average speed} = \frac{\text{Distance}}{\text{Time}}$$

estimate the peak speed of the runner. For instance, for the entire 100 meters the average speed is 10 meter/sec. But this is not peak speed because between the 50-

and 60-meter marks the average speed is $10/.95 = 10.5$ meter/sec. To convert to more familiar units of mph, simply divide by .447.

Meters	Seconds	Meters	Seconds
30	3.2	62	6.266
40	4.2	64	6.466
50	5.166	70	7.066
56	5.766	80	8.0
58	5.933	90	9.0
60	6.1	100	10.0

In what sense are all the times multiples of .033 seconds? Why would this be true? Because of the inaccuracies in the measurements, it is important to think about how reliable your estimate of peak speed is. For example, if the 60-meter mark were reached at 6.11 seconds instead of at 6.1 seconds, how much would your speed estimates change? Also, look at the pattern of average speeds as the runner moves from the 30-meter mark to the 100-meter mark: are there any unrealistic numbers? Taking all of this into account, how accurate do you think your estimate of peak speed is?

To improve our estimate we would need either a better video camera or a formula relating distance and time. The second half of this problem explores the latter (unrealistic) situation.

Suppose that the function $f(t) = 3t^2$ represents the distance covered in t seconds. For instance, after $t = 2$ seconds the runner has gone $f(2) = 12$ meters. The average speed between 1 and 2 seconds is

$$\frac{f(2) - f(1)}{2 - 1} = \frac{12 - 3}{2 - 1} = 9 \, \text{meter/sec}$$

What is the instantaneous speed at $t = 2$? We can get a better estimate than 9 meter/sec by computing the average speed between $t = 1.5$ and $t = 2$. Better still, compute the average speed between $t = 1.9$ and $t = 2$. Continue this process to estimate the speed at $t = 2$.

Further Study

The instantaneous rate of change, which you are asked to find above, is called the *derivative* and is an integral part of most calculus-based applications. We will give a more in-depth treatment of the derivative in Chapters 3 and 4.

2.2 Loss of Significance Errors

When conjecturing the value of a limit by using the evidence obtained from the TI-81/85 (or any other computational device, for that matter), we must always keep in mind that the numbers (and consequently also the graphs) obtained thereby are only approximate. Most of the time, this will cause us no serious trouble. The TI-81/85 carries out calculations to a very high degree of precision. Sometimes, however, the results of round-off errors in calculations are disastrous. In this section, we shall examine how and when such *loss of significance* errors occur. We'll also look at how to recognize these sometimes difficult to find computational errors and how to deal with their occurrence for a limited number of cases.

Example 1. A Flawed Calculation

Consider the limit

$$\lim_{x \to \infty} x[(x^2 + 4)^{1/2} - x]$$

Following the procedure worked out in the previous section, we draw several graphs to try to get a rough idea of the behavior of the function as x tends to infinity. Note that to enter this function, the brackets and braces displayed above must *all* be entered as parentheses:

$$x((x \wedge 2 + 4) \wedge (1/2) - x)$$

Figure 2.16a shows the initial TI-81 graph. Figure 2.16b was produced by resetting the x-range to Xmin = 0 and Xmax = 100 and the y-range to Ymin=−1 and Ymax=3.

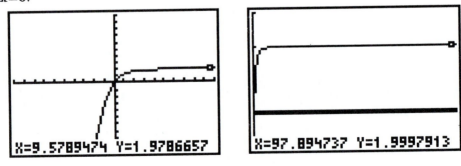

FIGURE 2.16a   FIGURE 2.16b

From the graphs (using the TRACE function), it seems that the function remains just about constant at around 2 as x goes further and further out to the right. Using FEVAL, we construct the following table:

x	$f(x)$	x	$f(x)$
10^1	1.980390272	10^5	2
10^2	1.99980004	10^6	2
10^3	1.999998	10^7	0
10^4	2	10^8	0

The last two values in the table should come as quite a surprise. They are inconsistent with what we expected from our examination of the graphs. They are also surprising since from $x = 10$ to $x = 10^6$, the corresponding function values seem to be homing in on 2. Then, all of a sudden, the values jump down to 0.

Could it be that we simply did not look at the graph for sufficiently large values of x? Certainly, this is always a possibility, since we're only drawing a large enough piece of the graph to try to get an idea of the limiting behavior as $x \to \infty$. However, this is not the case here. This is an example of a *loss of significance error*. The preceding table suggests that we look more carefully at what happens to the function between $x = 1,000,000$ and $x = 10,000,000$. We get the following from FEVAL on the TI-81 (the TI-85 will yield somewhat different results, with the loss of significance error occurring for larger values of x):

x	$f(x)$	x	$f(x)$
1,000,000	2	3,162,000	3.162
2,000,000	2	3,162,286	3.162286
2,500,000	2.5	3,162,287	0.0
3,000,000	3	3,162,288	−3.162288
3,100,000	3.1	3,162,289	0.0

This should strike you as rather strange. It would seem that the function is well-behaved, slowly approaching 2, as x approaches 2,000,000, but something unusual occurs beginning at about $x = 2,500,000$, with a very rapid change in the function values around $x = 3,162,288$. ∎

The reason for the unusual behavior witnessed in the last example boils down to how the TI-81/85 stores real numbers internally. Without getting too involved in the specifics of calculator arithmetic, it suffices to think of the calculator as storing real numbers in scientific notation. Thus, the real number 1234567 would be stored as 1.234567×10^6. The part of the number in front of the power of 10 is called the *mantissa* and the power of 10 is called the *exponent*. (So, here the mantissa is 1.234567 and the exponent is 6.)

Since no computing device has infinite memory, there are generally limitations on the size of the numbers which can be used by such a machine, as well as limits on the number of digits in the mantissa which are held internally by the machine. In the case of the TI-81, real numbers are permitted to have a mantissa between 1 and 9.999999999999 (13 digits) and an exponent between -99 and 99. For the TI-85, real numbers are permitted to have a mantissa between 1 and 9.9999999999999 (14 digits) with an exponent between -999 and 999. It should be noted that, while these both represent a high degree of accuracy (especially for a hand-held computing device), these limitations still present very real problems with accuracy in certain computations.

This suggests that real numbers are represented internally by only their first 13 or 14 significant digits. (This is referred to as *finite precision*.) Again, this is more than sufficient accuracy for most computations, but it does present an occasional problem. We will examine here what the consequences of such limited accuracy may be on the computation of limits. First, we look at several simple examples.

Example 2. Representation of Real Numbers in Finite Precision

On a 13-digit computer (such as the TI-81), $1/3$ is stored internally as
$$3.333333333333 \times 10^{-1}$$
On a 14-digit computer (such as the TI-85), $2/3$ is stored internally as
$$6.6666666666667 \times 10^{-1}$$

∎

In Example 3, we will see what happens if we subtract two numbers which differ only in the 15th significant digit.

Example 3. Arithmetic in Finite Precision

Notice that

$$1.00000000000002 \times 10^{17} - 1.00000000000001 \times 10^{17} = .00000000000001 \times 10^{17} = 1000$$

However, if the above operation is carried out on a machine with a 13- or 14-digit mantissa, both numbers on the left are represented by 1.0×10^{17} and consequently the difference between the two values is computed as 0. (Try this now on your TI-81/85.) Similarly,

$$1.00000000000006 \times 10^{20} - 1.00000000000004 \times 10^{20} = 2,000,000$$

In this case, if the calculation is carried out on a machine with a 14-digit mantissa, the first number on the left is represented by $1.0000000000001 \times 10^{20}$ and the second number as 1.0×10^{20} due to the limited accuracy and rounding. The difference between the two values is then computed as $.0000000000001 \times 10^{20}$ or 1.0×10^7 or 10,000,000. Once again, this is a serious error.   ∎

In both of the preceding calculations, we have a computed value which is grossly inaccurate, caused by a subtraction of two numbers whose significant digits are very close to one another. This type of error is referred to as a *loss of significant digits error* or simply a *loss of significance error*. These are subtle, but often disastrous, computational errors. Returning now to Example 1, we'll see that it was this type of computational error which caused the values after about $x = 2,000,000$ to be trashed.

Recall that the function under consideration was

$$f(x) = x[(x^2 + 4)^{1/2} - x]$$

Let's follow the computation for $x = 10,000,000$ one step at a time, as the calculator carries it out. First compute $(x^2 + 4)^{1/2}$ ($x = 1.0 \times 10^7$):

$$(x^2 + 4)^{1/2} = [(1 \times 10^7)^2 + 4]^{1/2} = [1 \times 10^{14} + 4]^{1/2}$$
$$= [1.00000000000004 \times 10^{14}]^{1/2}$$
$$= [1.0 \times 10^{14}]^{1/2}$$
$$= 1.0 \times 10^7$$

Recall that $1.00000000000004 \times 10^{14}$ is rounded off to 1.0×10^{14} because the calculator carries only 13 or 14 digits. Thus, the calculator gives

$$f(1 \times 10^7) = (1 \times 10^7)[1.0 \times 10^7 - 1.0 \times 10^7] = 0$$

Note that the real culprit here is not the rounding of 1.00000000000004 to 1, but the fact that this was followed by a subtraction. Additionally, notice that this is not a problem confined to the numerical computation of limits, but a problem typical of numerical computation, in general.

RULE OF THUMB: If at all possible, avoid subtractions, in order to avoid loss of significance errors. This can sometimes be accomplished by performing some algebraic manipulation of the function.

Returning once again to Example 1, notice that we may avoid the subtraction which seems to have been the cause of our problems, although in the process, we will complicate the expression for the function. Notice that

$$f(x) = x[(x^2 + 4)^{1/2} - x]$$

$$= x[(x^2 + 4)^{1/2} - x]\,\frac{(x^2 + 4)^{1/2} + x}{(x^2 + 4)^{1/2} + x}$$

$$= \frac{x[(x^2 + 4) - x^2]}{(x^2 + 4)^{1/2} + x} = \frac{4x}{(x^2 + 4)^{1/2} + x}$$

where, for $x > 0$, the last expression has no subtraction and hence also no loss of significance error. If we plot this last representation of the function, we get the same graphs as for the original representation of the function (Figures 2.16a - 2.16b). We now compute a table of values (from the TI-81, with somewhat different values from the TI-85 due to its slightly greater accuracy):

x	$f(x)$	x	$f(x)$
10^1	1.980390272	10^5	2.0
10^2	1.99980004	10^6	2.0
10^3	1.999998	10^7	2.0
10^4	1.99999998	10^8	2.0

and so on. This is more like the kind of progression of values which we had seen in earlier examples. From this evidence, together with a graph of the function over a large interval, we might now reasonably conjecture that

$$\lim_{x \to \infty} x[(x^2 + 4)^{1/2} - x] = 2$$

At this point, you should have a fairly good idea of how a loss of significance error can occur. In the next several examples, we shall pursue this a bit further.

Example 4. An Error Where Subtraction Is Not Explicitly Indicated

Consider $\lim\limits_{x \to -\infty} x[(x^2 + 4)^{1/2} + x]$. At first glance, you might think that since there's no subtraction explicitly indicated, there will be no loss of significance error. Upon closer examination, however, notice that since x is tending to *minus* infinity, an addition of two numbers of opposite sign (i.e., a subtraction) is taking place inside the brackets. Again, we can see the same peculiar behavior as that evident in Example 1. From Figure 2.17a (the initial TI-81 graph) and Figure 2.17b (from the TI-81, produced by resetting the x-range to Xmin$=-100$ and Xmax $= 0$ and the y-range to Ymin$=-3$ and Ymax $= 1$) it appears that the function tends to a value around -2 as $x \to -\infty$.

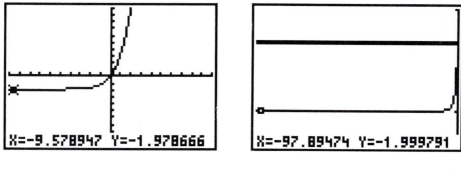

FIGURE 2.17a   FIGURE 2.17b

We obtain the following table from FEVAL (note that, as in Example 1, the function must be entered using only parentheses):

x	$f(x)$
-10^1	-1.980390272
-10^2	-1.99980004
-10^3	-1.999998
-10^4	-2.0
-10^5	-2.0
-10^6	-2.0
-10^7	0.0
-10^8	0.0

Again, the sudden change in values should appear suspicious. Upon closer examination, it should be clear that we do indeed have a loss of significance error (be sure you understand the graphical and numerical reasons for this conclusion). The remedy, as it was for Example 1, is to rewrite the expression.

$$f(x) = x[(x^2 + 4)^{1/2} + x]$$

$$= x[(x^2 + 4)^{1/2} + x] \frac{(x^2 + 4)^{1/2} - x}{(x^2 + 4)^{1/2} - x}$$

$$= \frac{x[(x^2 + 4) - x^2]}{(x^2 + 4)^{1/2} - x} = \frac{4x}{(x^2 + 4)^{1/2} - x}$$

Using this last expression, we construct the table:

x	$f(x)$	x	$f(x)$
-10^1	-1.980390272	-10^5	-2.0
-10^2	-1.99980004	-10^6	-2.0
-10^3	-1.999998	-10^7	-2.0
-10^4	-1.99999998	-10^8	-2.0

As in Example 3, the algebraic manipulation has eliminated the subtraction and, hence, the loss of significance error. We now conjecture that

$$\lim_{x \to -\infty} x[(x^2 + 4)^{1/2} + x] = -2$$

■

You will have noticed that the loss of significance errors in each of the last two examples occurred when the x-value reached about 10,000,000 in absolute value. Unfortunately, these errors do not occur only when dealing with numbers fairly large in absolute value. They can occur any time two nearly equal numbers are subtracted (or two numbers of nearly equal absolute values but opposite signs are added).

Example 5. A Loss of Significance Error Near $x = 0$

Consider $\lim_{x \to 0} \frac{1 - \cos(x)}{x^2}$. We use $\boxed{\text{ZOOM}}$ $\boxed{\text{BOX}}$ to zoom in on the portion of the initial TI-81 graph indicated in Figure 2.18a. The zoomed graph is shown in Figure 2.18b.

From the graphs, it appears that the function is approaching a value around .5 as $x \to 0$. Using FEVAL, we obtain the following table from the TI-81.

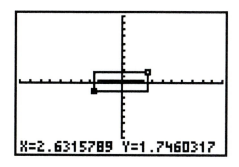
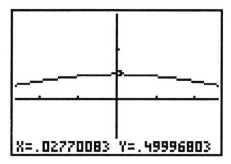

FIGURE 2.18a   FIGURE 2.18b

x	$f(x)$	x	$f(x)$
10^{-1}	0.4995834722	-10^{-1}	0.4995834722
10^{-2}	0.499995833	-10^{-2}	0.499995833
10^{-3}	0.5	-10^{-3}	0.5
10^{-4}	0.5	-10^{-4}	0.5
10^{-5}	0.5	-10^{-5}	0.5
10^{-6}	0.5	-10^{-6}	0.5
10^{-7}	0.0	-10^{-7}	0.0
10^{-8}	0.0	-10^{-8}	0.0

Notice that the y-values in the table take a sudden jump from .5 to 0.0. From the graphs in Figures 2.18a - 2.18b, we might reasonably expect a value near .5. This indicates that there may be a loss of significance error in the computation of $f(x)$. In this case, the value of $\cos(x)$ is very nearly equal to 1 when x is nearly 0. The subtraction of 1 and $\cos(x)$ (two very nearly equal values) then causes the error.

While from the first 5 entries in either column of the table and the graphs, we might reasonably conjecture that the value of the limit is .5 (in fact, this is correct) there is a broader question here. How can we reliably compute values of $f(x)$ for x close to 0? From the preceding, it should be clear that we cannot use the given representation of the function. Consider the following manipulation.

$$f(x) = \frac{1 - \cos x}{x^2} = \frac{1 - \cos x}{x^2} \cdot \frac{1 + \cos x}{1 + \cos x} = \frac{1 - (\cos x)^2}{x^2(1 + \cos x)} = \frac{(\sin x)^2}{x^2(1 + \cos x)}$$

Notice that, as in the previous examples, this last expression has no subtraction and hence, also will have no loss of significance error. We use this expression to construct the table:

x	$f(x)$	x	$f(x)$
10^{-1}	.4995834722	-10^{-1}	.4995834722
10^{-2}	.4999958333	-10^{-2}	.4999958333
10^{-3}	.4999999583	-10^{-3}	.4999999583
10^{-4}	.4999999996	-10^{-4}	.4999999996
10^{-5}	.5	-10^{-5}	.5
10^{-6}	.5	-10^{-6}	.5
10^{-7}	.5	-10^{-7}	.5

Once again, we have eliminated the loss of significance error by performing an elementary manipulation. We should be able to use the last expression above for accurate calculation of the values of $f(x)$ and we can, as well, make the conjecture:

$$\lim_{x \to 0} \frac{1 - \cos(x)}{x^2} = \frac{1}{2}$$

∎

We must point out that the method used to avoid the loss of significance error in the last three examples is not one which will work for all problems, or for most problems, for that matter. Again, these errors are often hard to find and somewhat tricky to fix. A more complete exposition of these is best deferred to a course in *numerical analysis*. Our main point in discussing these here is to make the student aware of them at an early point, since they will invariably be encountered in calculations. It is useful to be able to recognize when these errors occur and to know how to fix them in at least a limited number of cases.

Exercises 2.2

In exercises 1-6, conjecture the value of the limit. For what value of x does a loss of significance error appear?

1. $\lim\limits_{x \to \infty} x[\sqrt{4x^2 + 1} - 2x]$

2. $\lim\limits_{x \to -\infty} x[\sqrt{4x^2 + 1} + 2x]$

3. $\lim\limits_{x \to \infty} \sqrt{x}[\sqrt{x + 4} - \sqrt{x + 2}]$

4. $\lim\limits_{x \to \infty} x^2[\sqrt{x^4 + 8} - x^2]$

5. $\lim\limits_{x \to 0} \dfrac{x - \sin x}{x^3}$

6. $\lim\limits_{x \to 0} \dfrac{1 - \cos x^2}{x^4}$

In exercises 7-10, rework the indicated exercise after rewriting the function to reduce loss of significance error.

7. exercise 1   8. exercise 2

9. exercise 3   10. exercise 4

In exercises 11-13 (as well as the Exploratory Exercise) you will see what effect a small numerical error can have.

11. Compare $\lim\limits_{x \to 1} \dfrac{x^2 + x - 2}{x - 1}$ and $\lim\limits_{x \to 1} \dfrac{x^2 + x - 2.01}{x - 1}$.

12. Compare $\lim\limits_{x \to 2} \dfrac{x - 2}{x^2 - 4}$ and $\lim\limits_{x \to 2} \dfrac{x - 2}{x^2 - 4.01}$.

13. Compare $\sin \pi x$ and $\sin 3.14x$ for $x = 1$ (radian), $x = 10$, $x = 100$, and $x = 1000$.

EXPLORATORY EXERCISE

Introduction

The theory of *chaos* has been called the most interesting scientific development of the last 25 years. One of the fundamental principles of chaos is that small changes in numbers (such as those due to loss of significance errors) can have substantial effects on calculations. Physically, this concept is dramatized by the "Butterfly Effect." This states that the air stirred by a butterfly in China can create or disperse a hurricane in the South Atlantic two days later. Below we look at a basic example of chaos.

Problems

Set up a program called CHAOS (on the TI-81, it is convenient to store this as program PrgmC):

:PrgmC:CHAOS

:[Ans] [→] X

:X*(C-X)

:[Disp] [Ans]

The corresponding TI-85 program follows (we suggest you use the name CHAOS).

:[Ans] [→] x   :x*(C-x)

Start with C=2 (press 2 [→] C [ENTER]), press .5 [ENTER] and execute CHAOS several times. (Note: After the first execution of CHAOS, you can repeat the execution by simply pressing $\boxed{\text{ENTER}}$.) After a few times, the number 1 will be returned every time: we are "stuck" at 1. Now try C=2.5, put .5 on the stack and evaluate CHAOS until you get stuck at 1.5. This seems pretty tame: a small change in C produced a small change in output. But, try C=3.2 (again starting with .5 on the stack). This time, you will get stuck alternating between 2 different numbers. A small change in C produced a substantially different behavior.

Verify the following statements: with C=3.48 you get stuck on a 4-number cycle. With C=3.555 you get an 8-number cycle. With C=3.565 you get a 16-number cycle. With C=3.569 you get a 32-number cycle. With C=3.5697 you get a 64-number cycle. With C=3.57 you get chaos (no repetitions).

This information is summarized in the diagram below. For instance, the point (2,1) signifies that with C=2 you get stuck at 1; the point (2.5,1.5) signifies that with C=2.5 you get stuck at 1.5. The diagram divides into 2 branches and then 4 branches and then 8 branches and so on as the process develops 2-number cycles, 4-number cycles, 8-number cycles and so on. The places where the branches occur are called *bifurcation points*. Try to find them.

What happens with C=3.5? Since C=3.48 produces a 4-number cycle and C=3.555 produces an 8-number cycle, C=3.5 is a surprise! Try to identify where C=3.5 is on the diagram. Finally, what happens if you start with $x = 1$?

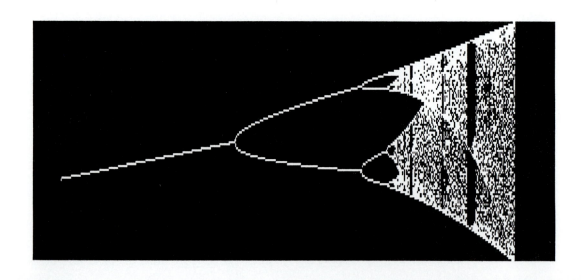

Further Study

Chaos is a very young scientific field, but there are several books out which are very well written and enjoyable. *Chaos* by James Gleick is the most general (it spent several weeks on the best seller list). The work we did above is the basis for current attempts to understand turbulence (see *Nonlinear Dynamics and Chaos* by Thompson and Stewart).

2.3 Exploring the Definition of Limit

We have now spent many pages discussing various aspects of the computation of limits. This may seem just a bit odd, in that we have never actually defined what a limit is. Oh, sure, we have an *idea* of what a limit is, but that's all. In section 2.1, we reminded the reader of the intuitive notion of a limit. Again, we say that

$$\lim_{x \to a} f(x) = L$$

if $f(x)$ gets closer and closer to L when x gets closer and closer to a.

We have so far been quite happy with this vague, although intuitive, description. For many purposes, this notion is quite sufficient. However, this needs to be made more precise. In doing so, we will begin to see how mathematical analysis (that branch of mathematics of which the calculus is the most elementary study) works.

Studying more advanced mathematics without an understanding of the precise definition of limit is somewhat like studying brain surgery without bothering with all that background work in chemistry and biology. A brain surgeon certainly doesn't seem to need these things to perform his or her job, but neither of the authors would consent to surgery by one who did not have a thorough understanding of these areas. Why not? In biology and medicine, it has only been through a careful examination of the microscopic world that a deeper understanding of our own macroscopic world has been achieved, and good surgeons need to understand what they are doing and why they are doing it. Mere technical proficiency is simply not enough. Likewise, in mathematical analysis, it is only through an understanding of the microscopic behavior of functions (here, the precise definition of limit) that real understanding of the mathematics will come about.

We begin with the careful examination of an elementary example.

Example 1. Analysis of a Limit of a Polynomial

Within the framework of our intuitive notion of limit, it's easy to believe that

$$\lim_{x \to 3}(2x + 5) = 11$$

The function is a polynomial and we have already observed that the limit of any polynomial is found by simply plugging in the value for x. But, why is that? What is it that this statement is trying to communicate? You might answer that as x gets closer and closer to 3, the quantity $(2x + 5)$ will get closer and closer to 11. But, this is rather vague. What do we mean when we say closer and closer?

A simple way to think about this is to say that we should be able to make $(2x + 5)$ as close as we might like to 11, simply by making x sufficiently close to 3. So, suppose that we want $(2x + 5)$ to be within, say, 1/2 of 11. Mathematically, this means that

$$-1/2 < (2x + 5) - 11 < 1/2$$

or adding 11,

$$11 - 1/2 < 2x + 5 < 11 + 1/2$$

For what values of x can we guarantee that this will be true? From the graph of $f(x) = (2x + 5)$, we can read off an answer. Since we are interested only in the behavior of the function near $x = 3$ and since we want the function values to lie between 10.5 and 11.5, we reset the x-range to be Xmin = 2 and Xmax = 4 and the y-range to be Ymin = 10.5 and Ymax = 11.5. See Figure 2.19 for the TI-81 graph. Using the TRACE function, it's easy to see that for x between about 2.76 and 3.25, the graph stays on the screen, i.e., $f(x)$ stays between 10.5 and 11.5.

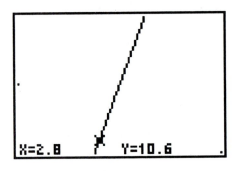

FIGURE 2.19

In this case, however, why not just solve the above inequality for x? We have

$$21/2 < 2x + 5 < 23/2$$

and subtracting 5,

$$11/2 < 2x < 13/2$$

so that

$$11/4 < x < 13/4$$

In particular, notice that this is equivalent to

$$3 - 1/4 < x < 3 + 1/4$$

More generally, just how close would you like $(2x + 5)$ to be to 11? We should be able to produce values of $2x + 5$ within *any* specified distance of 11. Pick some arbitrary distance and call it ϵ (*epsilon*, where $\epsilon > 0$). What range of values of x will guarantee that $(2x + 5)$ is within a distance of ϵ from 11? Figure 2.20 gives a graphical solution of this problem.

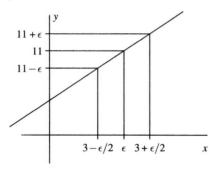

FIGURE 2.20

Again, we require that

$$-\epsilon < (2x + 5) - 11 < \epsilon$$

or

$$11 - \epsilon < 2x + 5 < 11 + \epsilon$$

Solving the inequality for x, we get

$$6 - \epsilon < 2x < 6 + \epsilon$$

or

$$3 - \epsilon/2 < x < 3 + \epsilon/2$$

and

$$-\epsilon/2 < x - 3 < \epsilon/2$$

We summarize this by saying that if we want $(2x + 5)$ to be within a distance of ϵ from 11, then x must be within $\epsilon/2$ of 3, i.e.,

$$|(2x + 5) - 11| < \epsilon \qquad \text{whenever} \qquad |x - 3| < \epsilon/2$$

■

Next, we consider this more precise notion of limit for an example where the function is undefined at the point in question.

Example 2. Analysis of a Limit of a Rational Function

Consider $\lim\limits_{x \to 2} \dfrac{3x^2 - 12}{x - 2} = 12$. As in Example 1, we would like to know how close x must be to 2, in order to guarantee that the function is within some arbitrary distance, ϵ $(\epsilon > 0)$, of 12. (See Figure 2.21 for a graphical representation of this problem.)

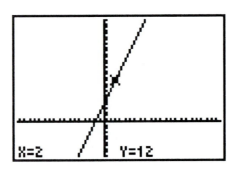

FIGURE 2.21

Notice that, in this case, the function is undefined at $x = 2$ and so we look for a number δ $(\delta > 0)$ such that if

$$0 < |x - 2| < \delta$$

then

$$\left| \frac{3x^2 - 12}{x - 2} - 12 \right| < \epsilon$$

(Note that we have $|x - 2| > 0$, so that $x \neq 2$, since the function is not defined at $x = 2$.) This last inequality corresponds to

$$12 - \epsilon < \frac{3x^2 - 12}{x - 2} < 12 + \epsilon$$

Since $x \neq 2$, we can factor and cancel, yielding

$$12 - \epsilon < \frac{3(x-2)(x+2)}{x-2} < 12 + \epsilon$$

or

$$12 - \epsilon < 3(x+2) < 12 + \epsilon$$

Solving this inequality for x, we get:

$$2 - \epsilon/3 < x < 2 + \epsilon/3$$

or

$$-\epsilon/3 < x - 2 < \epsilon/3$$

We can now see that choosing $\delta = \epsilon/3$ will do the job. That is, if

$$0 < |x - 2| < \epsilon/3$$

then

$$\left| \frac{3x^2 - 12}{x - 2} - 12 \right| < \epsilon$$

as desired.   ∎

This should give you a sufficiently clear idea of the process to make a general definition.

Definition (Precise Definition of Limit) For a function f defined in some open interval including a (except possibly at a itself), we say $\lim_{x \to a} f(x) = L$ if given any number $\epsilon > 0$, there is another number $\delta > 0$ such that whenever $0 < |x - a| < \delta$ we guarantee that $|f(x) - L| < \epsilon$.

We want to emphasize that this formal definition is not a completely new idea, but is simply a more precise expression of the intuitive notion which we have been using since the start of our discussion of limits. The difference is that we want to use this definition to carefully prove the conjectured value of a limit.

We have seen how to use the graphics and the computing power of the TI-81/85 to arrive at a conjecture for the value of a limit. But, how can we make use of the calculator in working with the above definition to prove that a limit has a certain value? In short, we cannot. However, it is of some value to explore the definition using the calculator. While we won't prove any theorems, we might gain some insight into what the definition is saying and how these δ's relate to the ϵ's.

First, as an illustration, let's look at a limit for which we can explicitly compute δ in terms of ϵ.

Example 3. Analysis of a Limit of a Quadratic Polynomial

Consider $\lim_{x \to 2}(x^2 - 1) = 3$. Given $\epsilon > 0$, we seek a $\delta > 0$ such that if $0 < |x - 2| < \delta$, then

$$|(x^2 - 1) - 3| < \epsilon$$

We can turn the problem around somewhat by assuming that for some $\delta > 0$, x lies within δ units of 2; i.e., $0 < |x - 2| < \delta$. Returning to the inequality, we have

$$|(x^2 - 1) - 3| = |x^2 - 4| = |x + 2| \cdot |x - 2|$$

If we further assume that x lies in the interval [1,3] (we are only interested in what happens near $x = 2$, anyway), we get that $|x + 2| \leq 5$, from which it follows that

$$|x^2 - 4| = |x + 2| \cdot |x - 2| \leq 5|x - 2|$$

We now require that

$$|(x^2 - 1) - 3| < 5|x - 2| < \epsilon$$

This occurs if and only if

$$|x - 2| < \epsilon/5$$

Thus, we can choose $\delta = \epsilon/5$ (as long as $\epsilon \leq 5$, so that x also stays in the interval [1,3]).

We should point out here that for most problems, finding the δ corresponding to a given ϵ is a very difficult task to accomplish algebraically. However, we can use the graphics of the TI-81/85 to gain insight into the relationship between δ and ϵ. First, we illustrate this for the present example.

Consider the choice $\epsilon = 1/2$. In this case, we are interested in what x-values near $x = 2$ will guarantee that the function values stay between $(3 - 1/2)$ and

$(3 + 1/2)$, i.e., between 2.5 and 3.5. In this case, we set the x-range to be [1,3] and the y-range to be [2.5,3.5] (i.e., press $\boxed{\text{RANGE}}$ and set Xmin = 1, Xmax = 3, Ymin = 2.5 and Ymax = 3.5). See Figure 2.22 for the TI-81 graph. Again using the TRACE function, you can easily observe that by keeping the x-values in the interval (1.9,2.1), the graph will stay on the screen (i.e., the y-values will stay in the desired interval). Notice that this set of x-values corresponds to those obtained from our previous analysis. (There, we had $|x - 1| < \delta$, where $\delta = \epsilon/5$, so that when $\epsilon = 1/2$, $\delta = 1/10$.) ∎

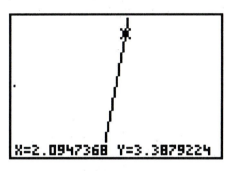

X=2.0947368 Y=3.3879224

FIGURE 2.22

We point out that an exploration of the definition of limit such as that exhibited in Example 3 is of most interest for problems where it is not obvious what the relationship between δ and ϵ might be. This is the case in Example 4.

Example 4. Exploring the Definition of Limit Where δ Is Unknown

Consider $\lim_{x \to 1} \cos(\pi x/2) = 0$. We would all certainly like to believe that this limit is correct. After all, $\cos(\pi/2) = 0$ and so this seems only fair. Also, any graph of $y = \cos(\pi x/2)$ will suggest that this should be true. Proving this is yet another matter. Given an $\epsilon > 0$, we look for a $\delta > 0$ such that

$$|\cos(\pi x/2) - 0| < \epsilon$$

whenever $0 < |x - 1| < \delta$. Finding δ in terms of an arbitrary ϵ is not easy. (Try this and see what we mean!) We can, however, use the graphics utilities of the TI-81/85 to experimentally find values of δ which work for any selected values of ϵ.

Let's start with the value $\epsilon = 1/2$. This means that we would like to know if we can find a $\delta > 0$ so that $0 < |x - 1| < \delta$ guarantees that

$$0 - 1/2 < \cos(\pi x/2) < 0 + 1/2$$

or

$$-1/2 < \cos(\pi x/2) < 1/2$$

Here, we are interested in what range of x-values (near $x = 1$) will guarantee that the function values will stay between $-1/2$ and $1/2$. Set the x-range to be $[0,2]$ and the y-range to be $[-.5, .5]$. (Press RANGE and set Xmin $= 0$, Xmax $= 2$, Ymin $= -.5$ and Ymax $= .5$.) See Figure 2.23a for the TI-85 graph. Using the TRACE function, you can see from the graph that if x is between about .68 and 1.32, the graph will stay on the screen, i.e., the y-values will fall in the desired range. This says that δ is approximately .32, here.

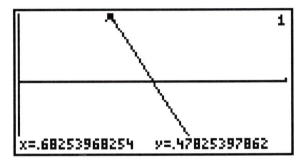

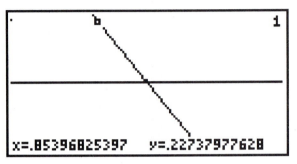

FIGURE 2.23a   FIGURE 2.23b

Next, we try $\epsilon = .25$. First, zoom in on the part of the graph of interest. Figure 2.23b was produced by setting the x-range to $[.6, 1.4]$ and the y-range to $[-.25, .25]$. Here, we want

$$-.25 < \cos(\pi x/2) < .25$$

In this case, notice that if x is between about .85 and 1.15, the graph stays on the screen and, hence, the y-values fall in the desired range. This gives us an approximate value for δ of .15.

Repeat this process for a few even smaller values of ϵ. We can continue this indefinitely. This is, of course, the whole idea of the definition of limit (i.e., that this process of finding δ 's corresponding to given ϵ 's can continue indefinitely). Again, while finding the δ 's for a few ϵ 's will not *prove* a conjecture as to the value of a limit, it should serve to illustrate the idea, as well as to provide evidence that our conjecture is correct. In the exercises, we will explore this idea further, both using the power of the TI-81/85 and by solving some problems by hand.   ∎

At this point, you should have a good understanding of what δ and ϵ represent.

In fact, you should be able to explain the reason for including each word and symbol in the definition of limit. One of the hallmarks of a good definition (and this definition is a great one, refined and improved over the last 150 years) is conciseness, and every piece of the definition is important. Yet, there is another crucial aspect of a definition that we have not addressed.

Find the flaw in the following definition of "dog": a dog is a four-legged mammal. The statement is certainly true enough, but it is not discriminating enough to be used as a definition. There are *lots* of four-legged mammals. A good definition must include characteristics which are true for dogs but *not* true for cats.

To return to mathematics, we must be sure that our definition of limit is not too general. Indeed, it is the ability of a definition to precisely delineate one category from another that is of paramount importance. In Example 5, we will see that a function which does *not* possess a limit at a point *does not* satisfy the properties of the definition. Then we may be satisfied with our definition.

Example 5. An Example Where the Limit Does Not Exist

Determine that $\lim\limits_{x \to 0} \dfrac{x^2 + x}{\sqrt{x^3 + 4x^2}}$ does not exist. To start our investigation, we use FEVAL to construct the following table of values.

x	$f(x)$	x	$f(x)$
10^{-1}	.5432512782	10^{-3}	.5004374492
10^{-2}	.5043699311	10^{-4}	.5000437495

From this table, we might conjecture a limit of .5. Note, however, that we have made a huge mistake in constructing the table (Where are the negative x's?) and we are getting misleading information. A sketch of the graph with the standard graphing window (see Figure 2.24) shows nothing unusual.

As we did in Example 4, we will examine the definition of limit graphically with $\epsilon = 1/2$. We would like to find a $\delta > 0$ so that $0 < |x| < \delta$ guarantees that

$$\frac{1}{2} - \frac{1}{2} < \frac{x^2 + x}{\sqrt{x^3 + 4x^2}} < \frac{1}{2} + \frac{1}{2}$$

or

$$0 < \frac{x^2 + x}{\sqrt{x^3 + 4x^2}} < 1$$

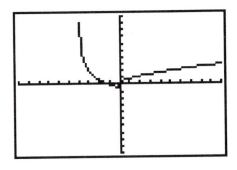

FIGURE 2.24

Set the x-range to be $[-.5, .5]$ and the y-range to be $[0,1]$ and redraw the graph (see Figure 2.25). This is a disturbing picture!

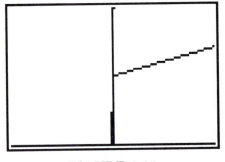

FIGURE 2.25

Using the TRACE command, we get y-values in $[0,1]$ only for positive x's. According to the definition, we must have y-values in $[0,1]$ *for all x's* between $-\delta$ and δ. Here, $\delta = .1$ does not work because for $x = -.005$ we have $-.1 < x < .1$ but $f(x)$ is *not* between 0 and 1 [in fact, $f(x) \approx -.4978$]. You should convince yourself that no matter how small we make δ, there is a value of x (specifically, a negative x) such that $0 < x < \delta$ but it is *not* true that $0 < f(x) < 1$. That is, there is no δ that makes the defining inequality true for $\epsilon = 1/2$.

There are several important points to make here. First, if we had used negative x's in our table of values we would never have made a bad limit conjecture. This mistake was compounded by an unclear (that is, insufficiently zoomed) graph in Figure 2.24. Second, the definition failed because we found a specific value of $\epsilon > 0$ for which no $\delta > 0$ would work. For the limit to exist, *all* positive numbers must have δ's. Third, we should make the technical point that what we actually showed in this example is that the limit is not $L = 1/2$. It is slightly more difficult to show that the limit could not equal any number. ∎

Exercises 2.3

In exercises 1-8, graphically find values of δ corresponding to $\epsilon = .1$ and $\epsilon = .05$.

1. $\lim\limits_{x \to 0} 3x + 5 = 5$

2. $\lim\limits_{x \to 2} 2x + 1 = 5$

3. $\lim\limits_{x \to 0} x^2 + 1 = 1$

4. $\lim\limits_{x \to -1} 2x^2 + 3 = 5$

5. $\lim\limits_{x \to 1} \sqrt{x + 3} = 2$

6. $\lim\limits_{x \to 2} \sqrt{x^3 + 1} = 3$

7. $\lim\limits_{x \to 1} \dfrac{x^2 + 3}{x} = 4$

8. $\lim\limits_{x \to 0} \cos x = 1$

In exercises 9-11 the function has the form $f(x) = kx$ for some constant k. Verify that $\delta = \epsilon/|k|$ works for $\lim\limits_{x \to 0} f(x) = 0$.

9. $f(x) = 3x$

10. $f(x) = -2x$

11. $f(x) = x/2$

12. Rework exercises 9-11 for $\lim\limits_{x \to 1} f(x)$. Would $\delta = \epsilon/|k|$ work for any c in $\lim\limits_{x \to c} f(x)$?

In exercises 13-16, verify graphically that the limits do not exist. Explain why there is no δ that works for $\epsilon = .1$. NOTE: to graph the function in exercise 13, enter $(X+1)(X\leq1)+X\wedge2(X>1)$ for Y1 (the inequality signs are in the Test menu). To get a clear picture, we suggest that you sketch a *disconnected* graph: on the TI-81, press MODE and choose Dot and on the TI-85, press FORMT and choose DrawDot .

13. $\lim\limits_{x \to 1} f(x)$ where $f(x) = \begin{cases} x + 1 & x \leq 1 \\ x^2 & x > 1 \end{cases}$

14. $\lim\limits_{x \to 2} f(x)$ where $f(x) = \begin{cases} x^3 - 4 & x \leq 2 \\ x^2 + 1 & x > 2 \end{cases}$

15. $\lim\limits_{x \to 1} \dfrac{4x}{(x - 1)^2}$

16. $\lim\limits_{x \to 0} \sin \dfrac{1}{x}$

17. State precisely what it means for a limit to not exist.

18. State precisely what is meant by $\lim\limits_{x \to c} f(x) = \infty$. Use $\lim\limits_{x \to 0} \dfrac{1}{x^2}$ as an example to guide your thinking.

19. A manufacturer of steel balls signs a contract to produce 2-lb balls. The customer allows a deviation of at most .02 lb. The radius and weight are related by $W = \dfrac{r^3}{4}$ (the radius is supposed to be 2 cm). How much can the radius deviate from 2 cm if the weight is to stay within the customer's specifications?

20. If the manufacturer in exercise 19 receives special orders with reduced tolerances of .01 lb and .005 lb, how much does the radius tolerance need to be reduced to?

EXPLORATORY EXERCISE

Introduction

In exercises 9-12, you found a general formula for δ in terms of ϵ for linear functions. Here, we will discover a similar formula for quadratic functions. In particular, we want to find a constant k such that $\delta = \epsilon/k$ works for $\lim_{x \to 0} ax^2 + bx + c = c$.

Problems

We will look at some examples before trying to generalize. For $f(x) = x^2 + 3x + 2$ zoom in on $(0,2)$ until the graph appears to be straight. Find two points on this curve and compute the slope m between these two points. Show that $k = m$ does not work but k slightly larger than m does work. Repeat this process for $f(x) = 4x^2 + 3x + 2$ and $f(x) = x^2 + x + 2$. Conjecture the slope of $f(x) = ax^2 + bx + c$ at $(0, c)$ and then conjecture a solution to our original problem.

Further Study

The work you have done above will be useful in several ways as you progress through calculus. You have probably discovered how to compute the *derivative* (slope) of a quadratic function. You have also come very close to proving an important result which states that if a function is differentiable at $x = a$ then it must be continuous at $x = a$.

CHAPTER
3

Differentiation

3.1 Construction of Tangent Lines

We are all familiar with the notion of a tangent line to a circle. This is a line which intersects the circle in exactly one point. Unfortunately, this idea does not generalize to all curves. We'll see that the tangent line to $y = x^3$ at $x = 1$ is given by $y = 3x - 2$, which intersects $y = x^3$ at $x = 1$ *and* $x = -2$. A standard way of defining the tangent line to the curve $y = f(x)$ at the point P is to consider a sequence of lines joining P with nearby points Q (these are called *secant lines*). As Q gets closer to P, the secant lines approach the tangent line (see Figure 3.1).

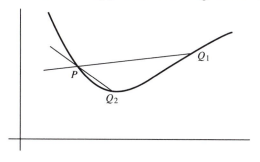

FIGURE 3.1

Notice that since we're already specifying the *point* of tangency, all we need to do to define the tangent line is to find its slope. We will do this shortly. But first, we want to come at this problem from a slightly different direction.

What is it that the tangent line is telling us about the graph of a function? One way to think about this is as follows. If you are walking along the graph of a function, the direction in which you are facing at any given point is along the tangent line. Further, if we zoom in enough on the graph, it should look fairly straight. The straight line that we then observe is an approximation to the tangent line. That should ring some bells. Your TI-81/85 is ideal for drawing graphs and zooming in on various points of interest.

Example 1. Using Graphics to Find an Approximate Tangent Line

Consider $f(x) = \dfrac{1}{2}x^3 - 1$, near $x = 1$. We will explore this idea of zooming in on the point $(1, -.5)$ until the graph looks fairly straight. We start with a graph of $y = f(x)$ using the default graphing parameters [enter the function in the "Y=" list and press $\boxed{\text{ZOOM}}$ 6 (Standard) on the TI-81 or press $\boxed{\text{ZOOM}}$ $\boxed{\text{ZSTD}}$ on the TI-85]. The initial TI-81 graph is shown in Figure 3.2.

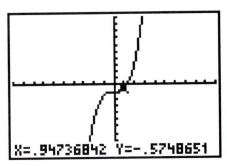

X=.94736842   Y=-.5748651

FIGURE 3.2

If you move the cursor over to the point $(1, -.5)$, you'll note that the graph does not look particularly straight nearby.

However, if we use the $\boxed{\text{ZOOM}}$ $\boxed{\text{BOX}}$ command (on either the TI-81 or the TI-85) we can see that the more that we zoom in on the point in question, the straighter the graph appears to be (see Figures 3.3a-3.3b for successively zoomed TI-81 graphs).

The endpoints of the portion of the graph appearing in Figure 3.3b are approximately $(.95844875, -.559773)$ and $(1.0812422, -.3679682)$ but yours will probably differ somewhat. The slope of the line joining these two points is then

$$m = \frac{y_2 - y_1}{x_2 - x_1} = \frac{-.3679682 - (-.559773)}{1.0812422 - .95844875} = 1.562011655$$

What does this slope represent? Well, it would seem to be an approximation

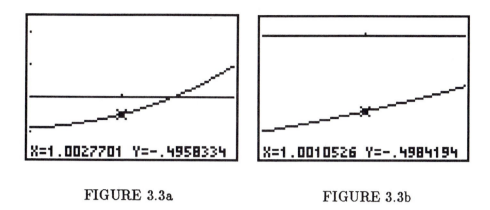

FIGURE 3.3a   FIGURE 3.3b

to the slope of the tangent line.

To observe what the line through $(1, -.5)$ with slope 1.562011655 looks like together with the original graph, plot them simultaneously, as follows:

1. Insert the equation of the approximate tangent line as Y_2 in the "Y=" list. [Move the cursor to Y_2 and enter: $Y_2 = 1.50364203952 * (X-1) - .5$.]

2. Press $\boxed{\text{ZOOM}}$ 6 (Standard) on the TI-81 or $\boxed{\text{ZOOM}}$ $\boxed{\text{ZSTD}}$ on the TI-85 to reset to the default graphics window and plot the two curves.

See Figure 3.4 for the TI-81 graph of this. From this, it is unclear if this is the tangent line that we are seeking. Actually, it's a secant line (i.e., a line joining two points on the graph) and acts as only a crude approximation to the tangent line.

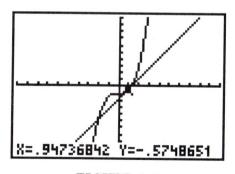

FIGURE 3.4

While we might be able to improve our approximation by zooming in even further on the point of tangency, there is a better way to think of this. You should keep in mind that the graphs produced by the TI-81/85 are at best fairly rough representations of the actual graphs. Thus, there's a built-in limit to how good an approximation of slope can be if it is derived from such an imperfect graph. ∎

You should expect to repeat this procedure (of zooming in on the point of tangency until the graph appears straight) more than once or twice. The process of zooming and writing down several points on paper and then keying those points back into the calculator (which is where they came from) for the slope calculation, is rather tedious. We can simplify this process with the following.

Once you have zoomed in sufficiently that the graph appears straight, do one more zoom. This time, press $\boxed{\text{ZOOM}}$ $\boxed{\text{BOX}}$ and for the two extreme corners of the box, use points *on the graph* (see Figure 3.5a and 3.5b).

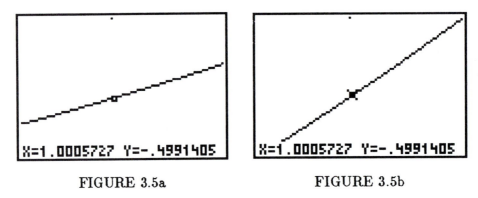

X=1.0005727 Y=-.4991405   X=1.0005727 Y=-.4991405

FIGURE 3.5a   FIGURE 3.5b

Although Figure 3.5b does not seem to be an improvement over Figure 3.5a, you will notice that the graph goes right into two diagonally opposed corners of the display. The coordinates of the corners (and hence also the coordinates of two points on this straight segment of the graph) are now stored in the memory of the TI-81/85 in the variables Xmax, Xmin, Ymax and Ymin. The following TI-81 program will then compute the slope:

:SLOPE   :($\boxed{\text{Ymax}}$ $-$ $\boxed{\text{Ymin}}$)/($\boxed{\text{Xmax}}$ $-$ $\boxed{\text{Xmin}}$)   : $\boxed{\text{Disp}}$ $\boxed{\text{Ans}}$

The Range variables Ymax, Ymin, Xmax and Xmin are found by pressing $\boxed{\text{VARS}}$ and moving the cursor over to the "RNG" option. To enter $\boxed{\text{Ymax}}$, press 5.

The corresponding TI-85 program is simply:

:(yMax−yMin)/(xMax−xMin)

In this case, you can type the variable names (paying careful attention to the upper and lower case letters) or access them from the Vars menu (press $\boxed{\text{VARS}}$ $\boxed{\text{MORE}}$ $\boxed{\text{MORE}}$ $\boxed{\text{RANGE}}$). When prompted for a name, we suggest that you enter SLOPE.

Note that since Ymax is always greater than Ymin and Xmax is always greater than Xmin, SLOPE always returns a positive value. This is the *absolute value* of the desired slope. If the line on the screen has negative slope, you will need to adjust the value manually. Even so, this is a much easier procedure than the one originally suggested.

In general, for a function f, if we want the slope of the tangent line to $y = f(x)$ at the point corresponding to $x = a$, then we should graphically zoom in on the vicinity of the point of tangency enough that the graph appears to be a straight line. We again want to make the point that appearances can be deceiving. However, the basic idea here is correct. We can accomplish the same thing algebraically, but without the inaccuracies inherent in the reliance on graphics alone.

Pick a point on the curve nearby the point of tangency, $(a, f(a))$, say $(a + h, f(a + h))$, for some small value h. If h is small enough [i.e., if $(a + h, f(a + h))$ is close enough to $(a, f(a))$] then the portion of the graph between these two points should appear fairly straight. In this case, the slope of the secant line joining these two points will approximate the slope of the tangent line. From the usual formula for slope, the slope of this secant line is

$$m_{sec} = \frac{y_2 - y_1}{x_2 - x_1} = \frac{f(a + h) - f(a)}{(a + h) - a} = \frac{f(a + h) - f(a)}{h}$$

Example 2. Computing the Slope of a Secant Line

Again, for $f(x) = \frac{1}{2}x^3 - 1$, compute the slope of the secant line joining the points corresponding to $x = 1$ and $x = 1.1$. We get

$$m_{sec} = \frac{f(1.1) - f(1)}{.1} = \frac{-.3345 - (-.5)}{.1} = 1.655$$

A better approximation to the slope of the tangent line might be obtained if we find the slope of the secant line joining $(1, -.5)$ and an even closer point, say the point corresponding to $x = 1.01$. We find

$$m_{sec} = \frac{f(1.01) - f(1)}{.01} = \frac{-.4848495 - (-.5)}{.01} = 1.51505$$

∎

There are several things to notice here. First, the slopes of the secant lines computed above are reasonably close to the approximation to the slope of the tangent line found graphically in Example 1. Second, it is reasonable to expect that the closer that the second point is chosen to the point of tangency, the closer the slope of the corresponding secant line should be to the slope of the tangent line. Rather than compute a long sequence of slopes manually, we suggest the following TI-81 program:

:MSEC   :Disp "ENTER H"   :Input H   :A → X
:Y₁ → Y   :A+H → X   :(Y₁ −Y)/H   :Disp Ans

Program Step	Explanation
:MSEC	Name the program.
:Disp "ENTER H"	Display a message on the home screen to prompt for a value of H.
:Input H	Accept a value for H.
:A → X	Store the value of A into X.
:Y₁ → Y	Compute $f(A)$ and store this in the variable Y.
:A+H → X	Store the value of A + H in the variable X.
:(Y₁ − Y) / H	Compute the slope of the secant line.
:Disp Ans	Display the result on the home screen.

The corresponding TI-85 program follows (when prompted for the program name, enter MSEC):

:Disp "ENTER H"   :Input H   :A → x
:y1 → Y   :A+H → x   :(y1−Y)/H

Before running MSEC (on either calculator), you will need to enter the function f into the "Y=" list as Y1 and store the x-coordinate of the point of tangency, x_0, in the variable A. For the present case, enter 1 → A ENTER .

When you execute MSEC , you will be prompted to enter a value for H. The program will then compute the slope of the secant line joining the points on the graph of $y = f(x)$ corresponding to $x = x_0$ and $x = x_0+$H. For example, using H=.1, we get the slope 1.655, as above. Note that if you press ENTER , the program will be immediately re-executed. This allows us to easily compute slopes of secant lines for a sequence of values of H getting closer and closer to 0. In this way, we can observe the limiting behavior of these slopes and, from this, conjecture the value of the slope of the tangent line.

H	MSEC	H	MSEC
10^{-1}	1.655	-10^{-1}	1.355
10^{-2}	1.51505	-10^{-2}	1.48505
10^{-3}	1.5015005	-10^{-3}	1.4985005
10^{-4}	1.500150005	-10^{-4}	1.499850005
10^{-5}	1.500015	-10^{-5}	1.499985
10^{-6}	1.5000015	-10^{-6}	1.4999985
10^{-7}	1.5	-10^{-7}	1.5

From this table, we see that the slopes of the secant lines seem to be getting closer and closer to 1.5 as H gets closer and closer to 0. Intuitively, then, it seems reasonable to conjecture that the slope of the tangent line is 1.5. To check that this conjecture is consistent with our geometric intuition about tangent lines, draw the graph of $y = f(x)$ together with that of the line through $(1, -.5)$ with slope 1.5. Again, enter

$$Y2 = 1.5 * (X - 1) - .5$$

in the "Y=" list and draw the graph using the default graphics window [press ZOOM 6 (Standard) on the TI-81 or ZOOM ZSTD on the TI-85]. The resulting TI-81 graph is shown in Figure 3.6.

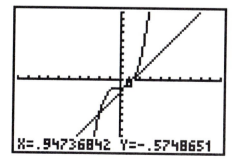

X=.94736842 Y=-.5748651

FIGURE 3.6

In the figure, it appears that this is indeed the tangent line that we're looking for. If you have a TI-85, you may have noticed that there is a built-in command for finding the slope of a tangent line and drawing its graph. After entering the function in the "y(x)=" list as y1 and drawing its graph (we suggest that you use $\boxed{\text{ZOOM}}$ $\boxed{\text{MORE}}$ $\boxed{\text{ZDECM}}$ to do this), press $\boxed{\text{EXIT}}$ (to return to the graphing menu) $\boxed{\text{MORE}}$ $\boxed{\text{MATH}}$ $\boxed{\text{MORE}}$ $\boxed{\text{MORE}}$ $\boxed{\text{TANLN}}$. The right and left arrow keys will now move the TRACE cursor along the graph. For the present case, move the cursor to the point $(1, -.5)$ and press $\boxed{\text{ENTER}}$ to draw the tangent line. The slope of the tangent line is displayed in the lower left corner of the screen. Note that the slope and graph match our slope conjecture from MSEC and the graph in Figure 3.6, respectively.

WARNING: If you take the value of H to be too small, the calculation of m_{sec} may be subject to a loss of significance error, as discussed in Chapter 2. For example, here a value of $H = 10^{-12}$ yields $m_{sec} = 1.5$, while $H = 10^{-13}$ produces the value $m_{sec} = 0.0$ on the TI-81, with similar results on the TI-85 (using $H = 10^{-13}$ and $H = 10^{-14}$, respectively, due to its greater accuracy).

Let's briefly review Example 2 to see how we might more precisely define the notion of tangent line. In the preceding, we computed the slopes of secant lines for a sequence of points getting closer and closer to the point of tangency. We observed that the limiting value of these slopes should be the slope of the tangent line. In general, we have:

Definition The slope of the tangent line to $y = f(x)$ at the point $(x_0, f(x_0))$ is

$$m_{tan} = \lim_{h \to 0} \frac{f(x_0 + h) - f(x_0)}{h}$$

provided the limit exists.

It should be stressed that when conjecturing the value of such limits numerically (e.g., with the TI-81/85) you should always compare the conjectured value with what you expect from the graph (by zooming in on the behavior near the point of tangency, as in Example 1) and further check that the line through the point of tangency with the conjectured slope looks like the tangent line when plotted simultaneously with the graph of the function.

Example 3. Conjecturing the Value of the Slope of the Tangent Line

Find the slope of the tangent line to the graph of $f(x) = x \sin(\pi x)$ at $x = 2$. First, we draw the graph of the function and zoom in repeatedly, until the graph appears fairly straight (see Figures 3.7a and 3.7b for the appropriate TI-85 graphs).

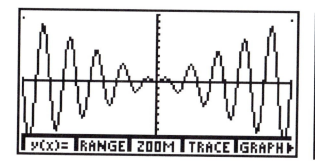

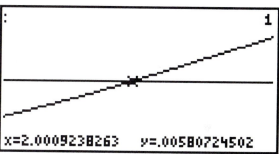

FIGURE 3.7a   FIGURE 3.7b

In Figure 3.7b the graph appears fairly straight, and if you use program SLOPE to compute the slope of the line joining two points on this segment of the graph, you should get a value around 6.2. (Try this now for yourself!)

Next, use the program MSEC to compute a sequence of values for m_{sec}. Note that you'll first need to store the new function f (as Y1 in the "Y=" list) and a new value for the x-coordinate of the point of tangency, x_0 (in the variable A).

Using MSEC, we generate the following table of values with the TI-81. The TI-85 will generate similar (but not identical) values.

H	MSEC	H	MSEC
10^{-1}	6.489356882	-10^{-1}	5.871322893
10^{-2}	6.313562575	-10^{-2}	6.250741057
10^{-3}	6.28631656	-10^{-3}	6.280033385
10^{-4}	6.283499356	-10^{-4}	6.282871037
10^{-5}	6.283216815	-10^{-5}	6.283153983
10^{-6}	6.283189142	-10^{-6}	6.283182858
10^{-7}	6.283180314	-10^{-7}	6.283179686

From the table, while the values may not look particularly nice, they seem to be getting closer and closer to a number around 6.2832. As further evidence, this is close to the value we expected from our use of the graphics. We should note that using even smaller values of H may not lead to progressively better approximations to the slope of the tangent line. Again, remember that these computations are highly subject to loss of significance errors. The results may be improved by using some of the hints found in section 2.2.

As a final test of the sensibility of our answer, we draw the graph of $y = f(x)$ with the suspected tangent line superimposed on it. That is, enter the function 6.2832 * (X − 2) in the "Y=" list as Y_2. See Figure 3.8a for the TI-85 graph with the default graphics parameters and Figure 3.8b for a zoomed-in graph. In both cases, the line drawn looks very much like the tangent line, as expected.

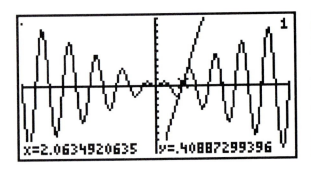

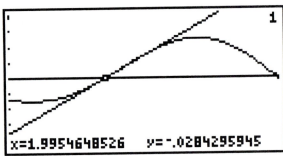

FIGURE 3.8a   FIGURE 3.8b

In this section, we have explored the notion of tangent line to the graph of a function and have seen how to compute approximations to the slope of the tangent line at a given point. We've also seen how to test our conjectured approximate slopes by using the graphics power of the TI-81/85. In section 3.2, we will examine a notion related to the slope of the tangent line and then see a more efficient way of computing approximations to these values.

Exercises 3.1

In exercises 1-8, use ZOOM, BOX and SLOPE to estimate the slope of the tangent line at the given point.

1. $f(x) = x^2 - 1$, $x = 1$

2. $f(x) = x^2 - 1$, $x = 2$

3. $f(x) = x^3 - x$, $x = 0$

4. $f(x) = x^3 - x$, $x = 1$

5. $f(x) = \sqrt{x^2 + 1}$, $x = 0$

6. $f(x) = \sqrt{x^2 + 1}$, $x = 1$

7. $f(x) = \sin x$, $x = 0$

8. $f(x) = \sin x$, $x = \pi$

In exercises 9-16, use MSEC to estimate the slope of the tangent line at the given point and compare to exercises 1-8.

9. $f(x) = x^2 - 1$, $x = 1$

10. $f(x) = x^2 - 1$, $x = 2$

11. $f(x) = x^3 - x$, $x = 0$

12. $f(x) = x^3 - x$, $x = 1$

13. $f(x) = \sqrt{x^2 + 1}$, $x = 0$

14. $f(x) = \sqrt{x^2 + 1}$, $x = 1$

15. $f(x) = \sin x$, $x = 0$

16. $f(x) = \sin x$, $x = \pi$

In exercises 17-18, use the graph and MSEC to determine that the slope of the tangent line does not exist.

17. $f(x) = |x|$, $x = 0$

18. $f(x) = (x^2)^{1/3}$, $x = 0$

In exercises 19-22, compute the slope of the tangent line by hand and compare your answers to those obtained in exercises 9-12.

19. $f(x) = x^2 - 1$, $x = 1$

20. $f(x) = x^2 - 1$, $x = 2$

21. $f(x) = x^3 - x$, $x = 0$

22. $f(x) = x^3 - x$, $x = 1$

In exercises 23-26, use the graph and MSEC to estimate the slope of the tangent line at $x = 0$, if it exists. HINT: The function in exercise 23 should be entered as $(x\wedge2-1)(x<0)+(x\wedge3-1)(x\geq0)$. The inequalities are located in the Test menu.

23. $f(x) = \begin{cases} x^2 - 1 & x < 0 \\ x^3 - 1 & x \geq 0 \end{cases}$

24. $f(x) = \begin{cases} x^2 - x & x < 0 \\ x^3 - x & x \geq 0 \end{cases}$

25. $f(x) = \begin{cases} x/2 & x < 0 \\ x/4 & x \geq 0 \end{cases}$

26. $f(x) = \begin{cases} x^2 - 1 & x < 0 \\ x^2 + 1 & x \geq 0 \end{cases}$

EXPLORATORY EXERCISE

Introduction

Progress in mathematics is often made through exploration with an eye towards finding patterns. It turns out that slopes of tangent lines are easily computed from the original function. That is, there are nice patterns for us to discover in the problems to follow.

Problems

Use MSEC to estimate the slope of the tangent line to $\sin x$ at $x = 0, .2, .4, .6, .8, ..., 3.0$. For instance, you should conclude that the slope of the tangent line at $x = 0$ is 1, the slope of the tangent line at $x = .2$ is approximately .98 (2 digits is sufficient), etc. This gives us 16 points on the slope function (more commonly called the *derivative* function) for $\sin x$: $(0,1)$, $(.2, .98)$, etc. What does this function look like? Since we do not yet have a formula for this function, we must be content with plotting the 16 known points. This is called a *scatter plot*. To get a scatter plot on the TI-81/85, first clear all functions out of the "Y=" list (or turn them off). Adjust the Range parameters: we suggest $Xmin = -1$, $Xmax = 4$, and $Ymin$ and $Ymax$ slightly smaller and larger, respectively, than the smallest and largest slopes you compute.

On the TI-81, then press $\boxed{\text{STAT}}$, highlight the "DATA" option and press 1 (Edit). Enter $x1 = 0$, $y1 = 1$, $x2 = .2$, $y2 = .98$, etc. When all 16 points have been entered, press $\boxed{\text{STAT}}$, highlight "DRAW" and press 2 (Scatter). On the TI-85, press $\boxed{\text{STAT}}$ $\boxed{\text{EDIT}}$ and name the x and y variables (we suggest "x1" and "y1"). Enter $x1 = 0$, $y1 = 1$, $x2 = .2$, $y2 = .98$, etc. When all 16 points have been entered, press $\boxed{\text{EDIT}}$ $\boxed{\text{DRAW}}$ $\boxed{\text{SCAT}}$.

Does this curve look familiar? By correctly identifying this curve you will find an easy rule for finding slopes of tangent lines to $y = \sin x$. Repeat the above for $y = \cos x$. Finally, estimate the slope of the tangent line to $y = e^x$ at $x = -1.5, -1.3, -1.1, ..., 1.5$. Then plot the data and identify the curve to find an easy rule for finding slopes of tangent lines to $y = e^x$.

Further Study

The rules found above are three of the basic *derivative* rules which you will use throughout the rest of your mathematics career. Rigorous derivations of these rules can be found in your calculus book.

3.2 Numerical Differentiation

There are two main concepts from which the notion of derivative follows. One is the notion of tangent line which we examined in section 3.1. The second concept is the notion of *velocity* .

We first want to briefly explore what is meant by the term velocity. Think of what you mean when you use the word. More importantly, ask yourself how you would *compute* velocity if asked. The first thought that comes to mind is probably the formula *distance = rate x time* which you first saw in high school algebra. If you want to know the rate (the velocity) then you need only divide the distance by the time. Simple, isn't it? It's also not generally correct, as we'll see.

For example, suppose that you are pulled over on the highway by a police officer. The officer steps up to your car and asks that dreaded question, "Do you know how fast you were going?" As a good student of mathematics, you might be tempted to answer something like, "Well, I've been driving for 3 years, 2 months, 7 days, 5 hours and about 45 minutes. I've kept very careful records and I can tell you that I've driven exactly 45,259.7 miles in that much time. Therefore, according to my calculations, I was going only

$$\frac{45,259.7 \, \text{miles}}{27,917.75 \, \text{hours}} = 1.62118 \, \text{mph}$$

Of course, the authors do not recommend that you use this argument if you're pulled over from the highway. To be sure, it's ridiculous and most police officers would be reaching for their handcuffs (or a straight jacket) by the time you finished explaining your calculation. But, *why* is this wrong? Certainly there's nothing wrong with the formula or the arithmetic. Well, the police officer might reasonably argue that during this 3-year period, you were not even in a car most of the time and, hence, the results are invalid.

Suppose that you think quick and substitute the following argument instead, "Officer, I left my house at exactly 6:17 pm tonight and by the time you pulled me over at 6:43 pm, I had driven exactly 17 miles. Now, that says that I was going only

$$\frac{17 \, \text{miles}}{26 \, \text{minutes}} \times \frac{60 \, \text{minutes}}{1 \, \text{hour}} = 39.23077 \, \text{mph}$$

well under the posted 45-mph speed limit."

What's wrong this time? Even assuming that you did not stop at all during the entire trip, this argument is still unconvincing. It should be clear that, although this is a much better estimate of your velocity than the 1.6 mph arrived at previously, you are still computing the velocity using too long of a time period. It's not hard to realize that since cars speed up and slow down (and can do so very quickly) we must compute the velocity using a much smaller time period. Well now, how small is small enough?

It's time for some answers. What we've been computing is what is usually referred to as *average velocity*. What we are really interested in (as well as what the police officer is interested in) is *instantaneous velocity*, the velocity at an instant in time. We must now see how this can be computed.

Example 1. Instantaneous Velocity

Suppose that you are driving in a straight line and that the distance that you've traveled at time t (measured in minutes) is given by the function

$$s(t) = \frac{1}{2}t^2 - \frac{1}{12}t^3 \qquad 0 \le t \le 4$$

Find the instantaneous velocity at $t = 2$ minutes. Here, we assume that $s(t)$ gives distance as measured in miles. As a starting point, we might compute the average velocity during the time interval $[0,2]$. We get

$$\text{Average velocity} = \frac{s(2) - s(0)}{2 - 0} = \frac{1.333333333 - 0}{2}$$

$$= .6666666667 \text{ miles/minute}$$

$$= 40 \text{ mph}$$

You might expect that we could improve our estimate by averaging over a smaller time interval. For example, on the interval $[1,2]$, we get

$$\text{Average velocity} = \frac{s(2) - s(1)}{2 - 1}$$

$$= 1.333333333 - .4166666667$$

$$= .9166666663 \text{ miles/minute}$$

$$= 54.99999998 \text{ mph}$$

Of course, we can continue this process indefinitely. The smaller we make the time interval, the better the approximation of the velocity should be. In general, on the time interval $[t_0 - h, t_0]$, the average velocity is given by

$$v_{\text{ave}} = \frac{s(t_0) - s(t_0 - h)}{t_0 - (t_0 - h)} = \frac{s(t_0) - s(t_0 - h)}{h}$$

Notice the similarity of this to the formula developed in section 3.1 for the slope of a secant line. We can use a program similar to MSEC to compute this value. We offer the following TI-81 program:

:VELOCITY :$\boxed{\text{Disp}}$ "ENTER H" :$\boxed{\text{Input}}$ H :A $\boxed{\rightarrow}$ X

:$\boxed{\text{Y}_1}$ $\boxed{\rightarrow}$ Y :A$-$H$\boxed{\rightarrow}$ X :(Y$-\boxed{\text{Y}_1}$)/H :$\boxed{\text{Disp}}$ $\boxed{\text{Ans}}$

Program Step	Explanation
:VELOCITY	Name the program.
:$\boxed{\text{Disp}}$ "ENTER H"	Display a prompt to enter the value of H.
:$\boxed{\text{Input}}$ H	Accept a value for H.
:A $\boxed{\rightarrow}$ X	Store the value of A in X.
:$\boxed{\text{Y}_1}$ $\boxed{\rightarrow}$ Y	Compute $f(A)$ and store this in Y.
:A$-$H$\boxed{\rightarrow}$ X	Store the value of A$-$H in X.
:(Y$-\boxed{\text{Y}_1}$) / H	Compute the average velocity on the interval [A$-$H,A].
:$\boxed{\text{Disp}}$ $\boxed{\text{Ans}}$	Output the result to the home screen.

The corresponding TI-85 program follows (when prompted for the program name, enter VELOCITY):

:$\boxed{\text{Disp}}$ "ENTER H" :$\boxed{\text{Input}}$ H :A $\boxed{\rightarrow}$ x

:y1$\boxed{\rightarrow}$ Y :A$-$H$\boxed{\rightarrow}$ x :(Y-y1)/H

Before you run either program, you will need to store a value for the time t_0 in the variable A and store the distance function S in the "Y=" list as Y1 (using x as the variable instead of t). For the present example, enter 2 $\boxed{\rightarrow}$ A $\boxed{\text{ENTER}}$.

When you execute VELOCITY, you will be prompted for a value of H. The program will then return the average velocity on the desired interval to the home screen. As usual, pressing $\boxed{\text{ENTER}}$ will immediately rerun the program. We can thus easily compute average velocities for a sequence of values of H, as in the following table (produced with the TI-81; the TI-85 produces similar although slightly different results).

H	AVE VEL	H	AVE VEL
1.0	0.9166666667	−1.0	0.9166666667
0.1	0.9991666667	−0.1	0.9991666667
0.01	0.9999916666	−0.01	0.9999916667
0.001	0.999999916	−0.001	0.999999917
0.0001	1.0	−0.0001	1.0
0.00001	1.0	−0.00001	1.0

Notice from the table that as we make the time interval over which we're averaging smaller and smaller, the average velocity seems to be getting closer and closer to 1 mile/minute (60 mph). This limiting value is what we mean by instantaneous velocity. We can now give a definition of velocity.   ∎

Definition If the position of an object traveling in a straight line at time t is given by the function $s(t)$ [i.e., $s(t)$ gives the location on a number line of the object], then the *instantaneous velocity* of the object at time t_0 is given by

$$\lim_{h \to 0} \frac{s(t_0) - s(t_0 - h)}{h}$$

or equivalently, by

$$\lim_{h \to 0} \frac{s(t_0 + h) - s(t_0)}{h}$$

Notice that this is precisely the same as the definition of the slope of the tangent line, except that, here, we have used the variable t instead of x. In particular, this says that the numerical computation of velocities will also be subject to loss of significance errors. Thus, taking H too small may result in a gross error in the computed average velocity and, hence, also in the conjectured value of the instantaneous velocity.

As we have noted, the limit in the preceding definition arises naturally in several different contexts. Actually, this limit is so common that we give it a name.

Definition The *derivative* of a function f is the function f' defined by

$$f'(x) = \lim_{h \to 0} \frac{f(x + h) - f(x)}{h}$$

The function f' is defined for every x for which the limit exists. If f' is defined at x_0, f is called *differentiable* at x_0.

You should note the relationship between derivative functions, tangent lines and velocities. The slope of the tangent line at the point $(x_0, f(x_0))$ is the *value* of the derivative function at $x = x_0, f'(x_0)$. Likewise, if $s(t)$ represents distance traveled along a straight line, then the velocity at time $t = t_0$ is $s'(t_0)$.

Your calculus text will spend a great deal of time developing rules for computing the derivatives of various common functions. These are extremely important, but it is not our intention to reproduce all of this material here. We refer the student to any standard calculus text for this discussion.

Next, we examine a method for getting improved approximations to the values of a derivative at a given point (i.e., improved over the computations done in section 3.1 and so far in this section). You should wonder why we would want to approximate something which we can compute exactly (symbolically) by hand. The explanation is that in practice we may have trouble computing derivatives symbolically. Sometimes, an expression is simply too complicated to make hand computation of the derivative practical. It may also happen that all we know about a function is a collection of data: measurements of the value of the function at a finite collection of points. For example, if you wanted to study the movement of a planet, you would not be handed a formula giving its position as a function of time. As with many real world problems, you would make observations of its position at a number of specific times and then try to determine, for example, the velocity from these discrete observations.

We could get very involved in a discussion of numerical differentiation, but such a discussion is best left to a text in numerical analysis (see for instance, Conte and deBoor, *Elementary Numerical Analysis*, 3rd edition). Instead, we give one reasonably good way of computing derivatives numerically. Recall that in order to approximate the value of a derivative at a point,

$$f'(x_0) = \lim_{h \to 0} \frac{f(x_0 + h) - f(x_0)}{h}$$

we have previously computed values of the difference quotient

$$\frac{f(x_0 + h) - f(x_0)}{h}$$

for values of h (both positive and negative) close to zero. For $h > 0$, this is called a *forward difference*, and for $h < 0$, this is a *backward difference*.

An alternative approach is to approximate $f'(x_0)$ by the *centered difference*

$$\frac{f(x_0 + h) - f(x_0 - h)}{2h}$$

for small values of $h > 0$. For reasons beyond the scope of this introductory discussion, it turns out that this centered difference is generally a better approximation to $f'(x_0)$ than either the forward or backward difference. Even so, we caution the reader that centered differences are highly subject to loss of significance errors as h gets close to zero, just as forward and backward differences are.

Example 2. Centered Difference Approximation to a Derivative

Approximate the value of $f'(x)$ at $x = 2$, for $f(x) = x \sin(\pi x)$. First, note that you can easily modify your MSEC or VELOCITY programs to compute centered differences [e.g., in MSEC by changing the line A $\boxed{\rightarrow}$ X to A$-$H $\boxed{\rightarrow}$ X and then dividing $(Y_1 - Y)$ by 2*H instead of by H]. You will also need to store a new value for X and enter the function f in the "Y=" list. We obtained the following table of values using the TI-81, where the DIFF column lists the values of the centered differences for the corresponding values of H. (Slightly different results are obtained with the TI-85.)

H	DIFF
10^{-1}	6.180339888
10^{-2}	6.282151816
10^{-3}	6.283174973
10^{-4}	6.283185197
10^{-5}	6.283185399
10^{-6}	6.283186
10^{-7}	6.28318
10^{-8}	6.2832
10^{-9}	6.284

We can compute the exact value of $f'(2)$, as follows. From the product rule and the chain rule,

$$f'(x) = \sin(\pi x) + \pi x \cos(\pi x)$$

Thus, $f'(2) = 2\pi = 6.28318530718$. Notice that this is very nearly the value computed in the preceding table with H = .0001. The later approximations in the table get progressively worse, although for the values of H displayed, they tend to stay "in the ballpark" of the exact value. Contrast this with the table of forward and backward differences (the MSEC values for $h > 0$ and $h < 0$, respectively) computed for this function in Example 3 of section 3.1. You should note that the centered difference values are more accurate for each given value of H. This is generally the case and, hence, it is usually better to use a centered difference approximation to a derivative than to use either a forward or backward difference approximation. However, you should note that the centered differences are still affected by loss of significance errors, as we had expected.   ∎

Unfortunately, there is no simple way of eliminating the loss of significance errors in these computations. Such a topic is found in a course in numerical analysis. Of course, for simple functions, you can always (and should) compute the derivatives exactly by hand. We also remind the reader that one can use the graphics discussed in section 3.1 as a check on the reasonableness of a centered difference approximation to a derivative.

Now that we have discussed the idea of centered differences, we should point out that the TI-81 and the TI-85 have built-in programs for computing centered differences. We note that the approximations are precisely those which you would obtain using the modified MSEC or VELOCITY program discussed above.

The TI-81 command NDeriv (for Numerical Derivative) is found in the Math menu: press $\boxed{\text{MATH}}$ and 8. The command can be used in several ways. To compute an approximate value for $f'(x)$, first store the desired value of x in the variable X and then enter the command NDeriv($f(x),h$), where $f(x)$ is an expression for the function of interest and h is a suitably small value. For example, entering $4 \boxed{\rightarrow}$ X and NDeriv($X^2 - 3, .001$) will compute the centered difference approximation to the derivative of $f(x) = x^2 - 3$ at $x = 4$, with $h = .001$.

Note that in practice, due to the likelihood of loss of significance errors, you should compute a sequence of approximate derivatives, using successively smaller values of h. You can do this by pressing the replay key $\boxed{\triangle}$ and then simply entering a new value for h. You can also use NDeriv in programs and with variables inserted for $f(x)$ and h, i.e., NDeriv(Y_1,H) will compute the centered difference approximation to the function stored under the name Y_1 (in the "Y=" list), at the value currently stored in the variable X, and with spacing h given by whatever value is currently stored in the variable H.

The syntax for the corresponding TI-85 command is somewhat different. The command nDer is located in the CALC menu (press $\boxed{\text{CALC}}$ $\boxed{\text{nDer}}$). Entering the command nDer($f(x)$,x) will compute the centered difference approximation to f' at the current value of the variable x, using spacing δ. Here, δ is a variable found in the Toler menu. Press $\boxed{\text{TOLER}}$ to find the current value of δ or to adjust the default value. For example, to compute the centered difference approximation to the derivative of $f(x) = x^2 - 2$ at $x = 3$, using spacing $\delta = .001$, first press $3\boxed{\rightarrow}$ x $\boxed{\text{ENTER}}$. Then check to see if the value of δ needs to be adjusted (.001 is the default) and enter: nDer($x^2 - 2$,x). The approximate derivative (6) should be displayed on the home screen.

As with the TI-81, the TI-85's nDer command may be used in programs and variables may be used in place of $f(x)$, i.e., nDer(y1,x) will compute an approximation to the derivative of whatever function is stored under the name y1 in the "y(x)=" list. Unfortunately, repeated calculation with a sequence of values of δ is possible only by adjusting the value of δ in the Toler menu and then retyping the nDer command.

While the nDer command on the TI-85 is quite useful, there is another feature of the TI-85 which is even more impressive. The TI-85 has a command for computing *exact* derivatives. The commands der1 and der2 (located in the CALC menu) will use the rules of differentiation (product rule, chain rule, etc.) to compute the first and second derivatives, respectively, *exactly* (i.e., symbolically). Although the derivative function itself is not displayed on the screen, it is computed internally and the value of the derivative function (at the current value of the specified variable) is then displayed on the home screen. The syntax of the two commands is the same as for nDer. For example, to calculate the exact derivative of $x\sin(x)$ at the current value of x, enter der1(x $\boxed{\text{SIN}}$ x,x). Likewise, to compute the second derivative of the same function, simply enter der2(x $\boxed{\text{SIN}}$ x,x).

Exercises 3.2

In exercises 1-6, use VELOCITY to estimate the instantaneous velocity at the given time.

1. $s(t) = \dfrac{2t^3}{t^2 + 1}$, $t_0 = 0$

2. $s(t) = \dfrac{2t^3}{t^2 + 1}$, $t_0 = 2$

3. $s(t) = \dfrac{2t^3}{t^2 + 1}$, $t_0 = 10$

4. $s(t) = \dfrac{t}{\sqrt{t^2 + 2}}$, $t_0 = 0$

5. $s(t) = \dfrac{t}{\sqrt{t^2 + 2}}$, $t_0 = 2$ 6. $s(t) = \dfrac{t}{\sqrt{t^2 + 2}}$, $t_0 = 10$

In exercises 7-12, use the TI-85 to calculate the derivative (if it exists) exactly at $x = 0$ and $x = 1$.

7. $f(x) = x^2 \sin x$ 8. $f(x) = x \sin x^2$ 9. $f(x) = \dfrac{x}{x^2 + 2}$

10. $f(x) = \dfrac{1 - x^2}{x^2 - 1}$ 11. $f(x) = (x \sin x)^2$ 12. $f(x) = \sqrt{x^2 + 4}$

In exercises 13-20, compare the centered difference, backward difference and forward difference at the given x_0 for $h=.1$, .01 and .001.

13. $f(x) = x \cos x$, $x_0 = 0$ 14. $f(x) = x \cos x$, $x_0 = 2$

15. $f(x) = \sqrt{x^2 + 1}$, $x_0 = 0$ 16. $f(x) = \sqrt{x^2 + 1}$, $x_0 = 2$

17. $f(x) = x^2 e^{-x^2}$, $x_0 = 0$ 18. $f(x) = x^2 e^{-x^2}$, $x_0 = 1$

19. $f(x) = \dfrac{\cos x}{x^2 + 1}$, $x_0 = 0$ 20. $f(x) = \dfrac{\cos x}{x^2 + 1}$, $x_0 = 1$

In exercises 21-22, compute centered differences with $h=.1$, .01 and .001 for the given function at $x_0 = 0$. In exercises 17-18 of section 3.1, you found that the derivative does not exist at $x_0 = 0$. What evidence does the centered difference give?

21. $f(x) = |x|$ 22. $f(x) = (x^2)^{1/3}$

23. With your TI-81/85 in degrees mode, store 0 in x and compute the derivative of $\sin x$ using der1 or nDer. To understand the answer, note that $\pi/180 \approx .01745$. Explain why there is a factor of $\pi/180$.

24. Suppose a car has an average speed of 60 mph in a 65-mph speed limit zone. As discussed in the text, the validity of the average speed depends on the length of the time inteval. For instance, a 60-mph average over 1 hour does not prove that the driver never exceeded the 65-mph speed limit. Suppose the car cannot speed up or slow down more than 1 mph in 1 second. For how long of a time interval does the 60-mph average speed guarantee that the driver did not break the speed limit?

In exercises 25-28, use the given data to estimate $f'(0)$.

25.

x:	-1	$-.6$	$-.2$	0.2	0.6	1
$f(x)$:	10	3.5	0.3	0.5	4.0	10

26.

x:	-1	$-.6$	$-.2$	0.2	0.6	1
$f(x)$:	2	1	.25	$-.15$	$-.25$	0

27.	x:	$-.3$	$-.2$	$-.1$	0	0.1	0.2	0.3
	$f(x)$:	$-.2$	$-.15$	$-.1$	0	0.1	$.25$	0.3
28.	x:	$-.3$	$-.2$	$-.1$	0	0.1	0.2	0.3
	$f(x)$:	2.0	1.4	1.1	1	1.2	1.5	2.2

In exercises 29-33, we preview an important result known as L'Hopital's rule. In exercises 29-32, start by graphing f/g and f'/g' simultaneously. Note that $f'/g' \neq f/g$ and $f'/g' \neq (f/g)'$. Then, use the graph and program FEVAL to compare $(f/g)(0)$ to $(f'/g')(0)$ and $\lim_{x \to 0} f/g$ to $\lim_{x \to 0} f'/g'$.

29. $f(x) = x^2$, $g(x) = \sin x$ 30. $f(x) = \sin 3x$, $g(x) = \sin 4x$

31. $f(x) = 1 - \cos x$, $g(x) = x^2$ 32. $f(x) = \sin x - x$, $g(x) = \tan x - x$

33. For exercises 31-32, compute $(f''/g'')(0)$ and compare to $\lim_{x \to 0} f/g$.

EXPLORATORY EXERCISE

Introduction

There are several ways of describing the characteristics of the graph of a function. In calculus, we typically use the properties of increasing or decreasing and concave up or down. As you look from left to right, if the graph goes up the function is *increasing* (for example, $y = x$). If the graph goes down the function is *decreasing* (for example, $y = -x$). If the graph curves up (like $y = x^2$) it is *concave up*. If the graph curves down (like $y = -x^2$) it is *concave down*.

Problems

In this exercise, we will discover the relationship between the values of f' and f'' and the properties of the graph of f. Start by simultaneously graphing $f(x) = x^3 - 3x^2 + 1$ and $f'(x) = 3x^2 - 6x$. How does the sign (+ or −) of f' relate to the properties (increasing/decreasing, concave up/down) of the graph of f? What do the zeros of f' correspond to? Repeat this for $f(x) = \sin x$ [$f'(x) = \cos x$]. Do any of your interpretations change for $f(x) = x^3$ [$f'(x) = 3x^2$]? Write down the relationship between f' and f in as much detail as possible. Now compare the graphs of $f(x) = x^3 - 3x^2 + 1$ and the second derivative $f''(x) = 6x - 6$. How does the sign (+ or −) of f'' relate to the graph of f? What does the zero of f'' correspond to? Repeat this for $f(x) = \sin x$ and $f(x) = x^3$. Write down the relationship between f'' and f in as much detail as possible.

Further Study

You have discovered the basic components of what are known as the First Derivative Test and Second Derivative Test, which you will see shortly in calculus.

3.3 Tangent Line Approximations

How does your calculator "know" that $\sin(1.2345678) = .9440056953$? Think about it. We all understand the processes of addition, subtraction, multiplication and division and are quite capable of performing even lengthy hand computations involving these operations (if we are really pressed to do so). Still, we very often use our calculators as a convenience. They save us the time and effort required for doing computations by hand. This is not the case for calculation of values of the trigonometric functions, exponentials, logarithms and even for computing fractional powers of real numbers. For these computations, we usually use our calculators because we know of no other way of finding these values. But, how does the calculator do it?

In this section, we would like to take a first step (although a very small step) toward understanding how certain kinds of approximations are made. We want to make the point early on that the technique which we will develop here is not terribly accurate. In fact, to be perfectly honest, we would describe these approximations as crude, at best. The values computed with the built-in functions of your TI-81/85 (or those of any other scientific calculator for that matter) will almost always be far better than those which we will develop here.

Why would we be interested in discussing such an inferior approximation? There are several reasons. First, the methods used internally in calculators are much too complicated to discuss here, while the method we develop will serve as an introduction to such approximations. Second, the method introduced here will guide us to the approximate solution of the more complicated problems found in the next section, problems which *cannot* be solved by the mere push of a button.

For a given function f, suppose that we need to find the value of f at the point x_1, where $f(x_1)$ is unknown. For example, $\cos(.5)$ is unknown, although we could use a calculator to approximate it (at least the authors don't know the value without using their calculators). The basic idea here is to find a value of x *near* x_1, say $x = x_0$, such that we already know the value of $f(x_0)$ exactly. If f is differentiable at x_0, draw in the tangent line to the graph of $y = f(x)$ at $x = x_0$ (see Figure 3.9).

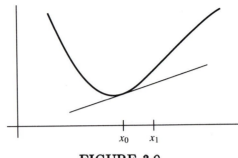

FIGURE 3.9

Keep in mind that the tangent line will "hug" the curve near the point of tangency (at $x = x_0$). This says that if x_0 is *close* to x_1, then the tangent line should still be close to the curve $y = f(x)$, at $x = x_1$. Examine Figure 3.9 to see that the y-values corresponding to $x = x_1$ on the curve and on the tangent line seem to be fairly close. To implement this idea, we need only find the equation of the tangent line. Since the slope of the tangent line is $f'(x_0)$ and the line passes through the point $(x_0, f(x_0))$, the equation is

$$\frac{y - f(x_0)}{x - x_0} = m_{\text{tan}} = f'(x_0)$$

or

$$y = f(x_0) + f'(x_0)(x - x_0)$$

The y-coordinate of the point on the line corresponding to $x = x_1$ (we will call this value y_1) is given by

$$y_1 = f(x_0) + f'(x_0)(x_1 - x_0)$$

To summarize, we are making the approximation

$$f(x_1) \approx f(x_0) + f'(x_0)(x_1 - x_0)$$

This is called the *tangent line* or *differential* approximation to $f(x_1)$.

Example 1. Tangent Line Approximation

Approximate the value of $\cos(.5)$. First, note that you can already get a highly accurate approximation from your TI-81/85: $\cos(.5) = .8775825619$. (Before doing this, make certain that your calculator is set to *radians* mode.)

The first step is to find the number closest to $x = .5$ at which we know the value of the cosine exactly. You should quickly realize that the closest such value is $x = \pi/6$. We know that $\cos(\pi/6) = \sqrt{3}/2$. Recall that for $f(x) = \cos(x)$, we have $f'(x) = -\sin(x)$. Thus, our approximation will be

$$\cos(.5) \approx \cos(\pi/6) - \sin(\pi/6)(.5 - \pi/6)$$
$$= \sqrt{3}/2 - .5(.5 - \pi/6) = .8778247916$$

Note that this is only a rough approximation to what we know to be the correct value, .8775825619.   ∎

Once again, we emphasize that this method routinely produces only mediocre approximations. What the reader should gain from this exposition (since the method itself is, in the present context, of minimal value) is an appreciation of how the method was developed. This will also serve as an introduction to the problems of section 3.4, for which more direct methods may not be available.

Exercises 3.3

In exercises 1-12, compute the tangent line to $f(x)$ at $x = x_0$ and use it to approximate $f(x_1)$ for the given values of x_1. Compare to the exact values.

1. $f(x) = x^2$, $x_0 = 1$; $x_1 = -1, 0, 2, 3$
2. $f(x) = x^3$, $x_0 = 1$; $x_1 = -1, 0, 2, 3$
3. $f(x) = \sin x$, $x_0 = 0$; $x_1 = -\pi/4, -\pi/6, \pi/6, \pi/4$
4. $f(x) = \sin x^2$, $x_0 = 0$; $x_1 = -\sqrt{\pi/4}, -\sqrt{\pi/6}, \sqrt{\pi/6}, \sqrt{\pi/4}$
5. $f(x) = \sqrt{x + 1}$, $x_0 = 0$; $x_1 = 1, 2, 3, 4$
6. $f(x) = \sqrt{x + 1}$, $x_0 = 3$; $x_1 = 1, 2, 4, 5$
7. $f(x) = (x^2 + 1)^{1/3}$, $x_0 = 0$; $x_1 = .5, 1, 1.5, 2$
8. $f(x) = (x^2 + 4)^{1/3}$, $x_0 = 2$; $x_1 = 1, 1.5, 2.5, 3$
9. $f(x) = \cos x$, $x_0 = 0$; $x_1 = -\pi/4, -\pi/6, \pi/6, \pi/4$
10. $f(x) = \cos x$, $x_0 = \pi/2$; $x_1 = \pi/4, \pi/3, 2\pi/3, \pi$
11. $f(x) = e^x$, $x_0 = 0$; $x_1 = -2, -1, 1, 2$
12. $f(x) = e^x$, $x_0 = 2$; $x_1 = 0, 1, 3, 4$

In exercises 13-18, find a tangent line approximation for the given value as was done in Example 1.

13. $\cos(2)$ 14. $\cos(1.5)$ 15. $\sqrt{4.2}$

16. $\sqrt[3]{8.4}$ 17. $\tan(1)$ 18. $\sin(.1)$

19. Suppose a person weighs P lb at sea level. At x ft above sea level, the person will weigh $W = \dfrac{PR^2}{(R+x)^2}$ lb, where $R \approx 21,120,000$ ft is the radius of the earth. Compute the tangent line approximation to W at $x_0 = 0$. In many applications, weight is considered to be constant. Why is this a reasonable assumption for applications near the surface of the earth?

EXPLORATORY EXERCISE

Introduction

We saw graphically that the tangent line gives a good approximation of a graph near the point of tangency. We can approach the idea of approximating a graph in a different way. Suppose we want a linear approximation of $f(x)$ near $x = x_0$. Since a line is determined by a point and the slope, the most we can demand is for the line to pass through $(x_0, f(x_0))$ and have slope $f'(x_0)$. The line with these properties is the tangent line $y = f'(x_0)(x - x_0) + f(x_0)$. We extend this idea below.

Problems

What is the best quadratic approximation of $f(x)$ near x_0? To simplify matters, take $x_0 = 0$. Quadratic functions have the form $Q(x) = ax^2 + bx + c$. Since there are 3 constants, we can make 3 demands. As above, we want $Q(x)$ to pass through $(0, f(0))$ and have slope $f'(0)$ at $x = 0$. Explain why this means $Q(0) = f(0)$ and $Q'(0) = f'(0)$. Our last demand is $Q''(0) = f''(0)$ (graphically, this forces Q to curve in the same direction as f). Show that these requirements are satisfied with $Q(x) = \dfrac{1}{2}f''(0)x^2 + f'(0)x + f(0)$. Repeat exercises 3, 5, 7 and 11 using $Q(x)$. Compare the accuracies of $Q(x)$ and the tangent line.

Further Study

$Q(x)$ is also known as the *second degree Taylor polynomial* of $f(x)$ centered at $x = 0$. By looking at higher-degree polynomials and demanding that higher-order derivatives match, you can derive higher-degree Taylor polynomials. We will study these in section 6.3.

3.4 Euler's Method

Many important phenomena in science and engineering are modeled by *differential equations*. In short, a differential equation is any equation involving the derivative or rate of change of an unknown function. For example,

$$y' - 2xy = x^2 \sin x$$

$$\frac{dy}{dx} + \cos y = (x^3 - 7x)\tan(\frac{x}{2})$$

$$2y(y')^3 - 3\cos x \sin y = 2x^2 - 3x + 7$$

are all differential equations for the unknown y as a function of x. Since these all involve only first derivatives, they are called *first-order* equations. The objective in solving these is to find a function y (a *solution*) which, when substituted into the equation, produces an identity (i.e., which satisfies the equation).

Example 1. Radioactive Decay

Physicists and chemists have long observed that radioactive substances decay at a rate directly proportional to the amount present. That is, if $Q(t)$ represents the amount of a certain radioactive substance present at time t, the rate of change of the amount with respect to time is

$$Q'(t) = kQ(t)$$

where k is the constant of proportionality (the *decay constant* which is known for any given radioactive substance). Note that, for any constant A, if

$$Q(t) = Ae^{kt}$$

then

$$Q'(t) = kAe^{kt} = kQ(t)$$

That is, $Q(t) = Ae^{kt}$ is a solution of the differential equation. In fact, since this function is a solution for every choice of the constant A, we have found infinitely many solutions, a different one for each value of A. A TI-85 graph of several of these is shown, for the case $k = -1$, in Figure 3.10. ∎

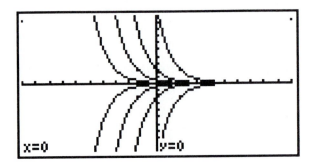

FIGURE 3.10

How are we to distinguish among these infinitely many solutions? If the question were left up to a 5-year-old, he or she would select one by *pointing* to the one they wanted, i.e., by placing their finger on a specific point, say (x_0, y_0), of the desired curve. This is exactly what is done in practice. We specify that the solution curve we are looking for should pass through the point (x_0, y_0). That is, if the solution is $y = y(x)$, we require that

$$y(x_0) = y_0$$

This is called an *initial condition* since, in applications, x_0 often represents an initial time. A differential equation together with an initial condition is called an *initial value problem*.

For the solution of Example 1, if Q_0 is the amount of the radioactive substance present at time $t = 0$, we have the initial condition

$$Q(0) = Q_0$$

from which it follows that

$$Q_0 = Q(0) = Ae^0 = A$$

Thus, the solution of the initial value problem (i.e., the amount of the substance present at time t) is

$$Q(t) = Q_0 e^{kt}$$

While we were able to guess the solution of the differential equation in Example 1, most problems are not so simple. There are many techniques available for solving various classes of first-order differential equations, but the study of these is best left to a concentrated course on differential equations. (There are many excellent references available; see, for example, *Elementary Differential Equations and Boundary Value Problems*, 5th edition, Boyce and DiPrima.) There are still many problems that are not easily solved directly. For such problems, we need to approximate the solution numerically. This is the purpose of the remainder of our discussion in this section.

Using the ideas developed in the last section, we can see how to approximate the solution of a first-order initial value problem. Consider the general case:

$$y' = f(x, y) \qquad y(x_0) = y_0$$

Suppose that the solution is $y = \phi(x)$. Then the differential equation gives us the slope of the tangent line to the graph of $y = \phi(x)$ at any given point (x, y) on the graph. In particular, from the initial condition, (x_0, y_0) is a point on the graph of the solution curve (you might think of this as the starting point). The slope of the tangent line at (x_0, y_0) is then given by the differential equation as $f(x_0, y_0)$. In Figure 3.11, you will see a typical solution curve, together with the tangent line at the initial point (x_0, y_0).

Note that if x_1 is close to x_0, then we might be able to use the tangent line to approximate the value of the solution at x_1.

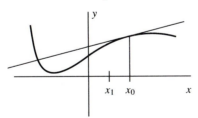

FIGURE 3.11

The equation of the tangent line is

$$m_{\tan} = \phi'(x_0) = f(x_0, y_0) = \frac{y - y_0}{x - x_0}$$

Like the tangent line approximation developed in section 3.3, if we follow the tangent line to the point corresponding to $x = x_1$, then the y-value at that point (call it y_1) should approximate the value of the solution there. That is,

$$\phi(x_1) \approx y_1 = y_0 + f(x_0, y_0)(x_1 - x_0)$$

We now have an approximation to the value of the solution at some $x = x_1$. However, we cannot just carelessly use this formula to find approximate values of the solution at *any* point. Again, you can clearly see from Figure 3.11 that such an approximation should be valid only when x_1 is close to x_0. Fortunately, there is another way to think about this problem.

You should note that, unlike the tangent line approximations we sought in the last section, when we look for the solution of an initial value problem, we are looking for a function, or at least for the value of that function at a number of points. In practice, we seek the solution of such a problem on an interval $[a, b]$, where a is usually the initial value x_0. When we look for approximate solutions numerically, we usually look for approximate values of the solution function at a finite number of points. For the sake of simplicity, we choose equally spaced points, starting at $x_0 : x_1, x_2, x_3, \ldots$, where

$$|x_{j+1} - x_j| = h \qquad j = 0, 1, 2, \ldots$$

and h is the *step size*. (See Figure 3.12 for an illustration of this partition of the interval.)

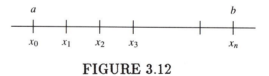

FIGURE 3.12

We use the tangent line approximation at x_1, i.e.,

$$\phi(x_1) \approx y_1 = y_0 + f(x_0, y_0)(x_1 - x_0)$$
$$= y_0 + hf(x_0, y_0)$$

Next, we want to find an approximation to $\phi(x_2)$. Certainly, if we knew the equation of the tangent line at the current point $(x_1, \phi(x_1))$, we could proceed as above by following this line to the point corresponding to $x = x_2$. The problem with this is that we don't even know the y-coordinate, $\phi(x_1)$. However, we do have an approximation for it, produced in the last step. Using the tangent line approximation, we can say that

$$\phi(x_2) \approx \phi(x_1) + f(x_1, \phi(x_1))(x_2 - x_1)$$
$$\approx y_1 + hf(x_1, y_1)$$

where in the second line we have approximated the slope of the tangent line, $f(x_1, \phi(x_1))$, with $f(x_1, y_1)$. [This would seem to be reasonable if y_1 is close to $\phi(x_1)$.] Continuing with this process, to get the approximation at x_3, we follow the approximate tangent line at the approximate point (x_2, y_2) to the point corresponding to $x = x_3$, and so on. We have the sequence of approximate values,

$$\phi(x_{n+1}) \approx y_{n+1} = y_n + h f(x_n, y_n) \qquad n = 0, 1, 2, \ldots$$

This is called *Euler's method* for approximating the solution to an initial value problem. See Figure 3.13 for an illustration of an exact solution versus the approximate solution derived from Euler's method (the broken line graph represents the approximate solution).

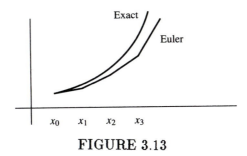

FIGURE 3.13

Example 2. Euler's Method for an Initial Value Problem

Consider the very simple initial value problem

$$y' = 2\frac{y}{x} \qquad y(1) = 2$$

First, we note that the exact solution of this problem is $y = \phi(x) = 2x^2$. [Verify this for yourself by differentiating $\phi(x)$ and plugging it into the equation and initial condition. We mention the exact solution here only so that we have some basis for comparison with our approximate solution.] If we did not know the exact solution (which is usually the case), we could try to approximate it using the Euler's method approximation given above, as follows.

Starting at the initial point (1,2), and using a step size of $h = .1$, we get

$$\phi(1.1) \approx y_1 = y_0 + hf(x_0, y_0)$$

$$= y_0 + h\frac{2y_0}{x_0} = 2 + .1\frac{4}{1} = 2.4$$

$$\phi(1.2) \approx y_2 = y_1 + h\frac{2y_1}{x_1} = 2.4 + .1(4.8/1.1)$$

$$= 2.4 + .436363636 = 2.836363636$$

$$\phi(1.3) \approx y_3 = y_2 + h\frac{2y_2}{x_2} = 3.309090909$$

and so on. Notice that these calculations are very repetitive. The programs which follow will make it simple to compute further values.   ∎

PROGRAM NOTE: Although the function f is a function of two variables, we can still enter this into the "Y=" list on the TI-81/85. In this case, simply enter 2*Y/x for Y1.

We suggest the following TI-81 program for Euler's method:

:EULER   :(Y+H*$\boxed{Y_1}$)$\boxed{\rightarrow}$ Y   :X+H$\boxed{\rightarrow}$ X   :$\boxed{\text{Disp}}$ X   :$\boxed{\text{Disp}}$ Y

Program Step	Explanation
:EULER	Name the program.
:(Y+H*$\boxed{Y_1}$)$\boxed{\rightarrow}$ Y	Compute the new y-value.
:X+H$\boxed{\rightarrow}$ X	Compute the new x-value.
:$\boxed{\text{Disp}}$ X	Display the new x-value on the home screen.
:$\boxed{\text{Disp}}$ Y	Display the new y-value on the home screen.

Here, we have used $\boxed{Y_1}$ to refer to the function of two variables stored in the "Y=" list which computes values of $f(x, y)$.

The corresponding TI-85 program follows (when prompted for a name, enter EULER):

:(Y+H*y1)$\boxed{\rightarrow}$ Y   :x+H$\boxed{\rightarrow}$ x   :$\boxed{\text{Disp}}$ x   :Y

To run this program, you'll first need to store a value for the variable H. In this case, press: .1 $\boxed{\rightarrow}$ H $\boxed{\text{ENTER}}$. Further, the program requires you to store the initial values x_0 and y_0 in the variables x and Y, respectively. Executing the $\boxed{\text{EULER}}$ program will then compute the approximate solution at $x = x_1 = x_0 + h$ and return the values x_1 and y_1 to the last two lines of the home screen. Continuing to press $\boxed{\text{ENTER}}$ will compute approximations at further points. Using $h = .1$, we constructed the following table.

We have also listed the value of the exact solution, since this was available to us. This allows us to determine the accuracy.

X	Approximate	Exact
1.1	2.4	2.42
1.2	2.836363636	2.88
1.3	3.309090909	3.38
1.4	3.818181818	3.92
1.5	4.363636364	4.5
1.6	4.945454545	5.12
1.7	5.563636364	5.78
1.8	6.218181818	6.48
1.9	6.909090909	7.22
2.0	7.636363636	8.0

Note that the approximate solution given here leaves much to be desired. In particular, notice that the further x gets away from the initial value of 1.0, the worse the approximation tends to get. This is characteristic of Euler's method and other numerical methods for solving initial value problems. While we can improve the results somewhat by using a smaller value of h, we cannot make h too small without facing the effects of loss of significance errors. Furthermore, the smaller the value of h is, the larger the number of steps will be that are required to reach a given x-value. This is illustrated in the following table.

H	Approximation	Error	Steps
.1	7.636363636	.3636363636	10
.05	7.80952381	.1904761905	20
.025	7.902439024	.0975609756	40
.0125	7.950617284	.049382716	80
.00625	7.97515528	.0248447205	160

Here, we compare the number of steps and the accuracy of the approximation at $x = 2.0$ for several values of h, again for the initial value problem in Example 2. The error listed is the absolute value of the difference between the exact solution and the approximate solution at $x = 2.0$.

Notice that as the step size decreases and the amount of effort increases (i.e., the number of steps increases), the accuracy of our approximation improves, but not dramatically. (Try this for yourself and see how quickly you get tired.) We could certainly write an automated program, but this will not correct the fundamental inefficiency of this method. The trouble with Euler's method is that too small a step size and, hence, too many calculations are required to obtain a reasonable approximation. In the exercises, we will explore a related but somewhat more efficient method for approximating the solutions of differential equations. For the moment, we should be content that we've developed a method which provides some minimal accuracy for solving problems which have no other apparent means of solution.

Exercises 3.4

In exercises 1-4, show that $\phi(x)$ is a solution of the differential equation for any c.

1. $\phi(x) = cx^2$, $\quad xy' - 2y = 0$ 2. $\phi(x) = ce^{2x}$, $\quad y' - 2y = 0$
3. $\phi(x) = x^2 + c$, $\quad y'' + y' = 2x + 2$ 4. $\phi(x) = c \sin x$, $\quad y'' + y = 0$

In exercises 5-8, find the value of c such that ϕ satisfies the given initial condition.

5. $\phi(x) = cx^2$, $\quad y(1) = 3$ 6. $\phi(x) = ce^{2x}$, $\quad y(0) = 5$
7. $\phi(x) = x^2 + c$, $\quad y(1) = 3$ 8. $\phi(x) = c \sin x$, $\quad y(0) = 2$

In exercises 9-16, use EULER with $h=.1$ to estimate the solution at $x = 1$.

9. $y' = 2xy$, $\quad y(0) = 1$ 10. $y' = \dfrac{x + 1}{y}$, $\quad y(0) = 1$

11. $y' = 3x - y$, $\quad y(0) = 2$ 12. $y' = y^2 - x$, $\quad y(0) = 2$
13. $y' = (x + y)^2$, $\quad y(0) = 1$ 14. $y' = x^2 + y^2$, $\quad y(0) = 1$
15. $(x + 1)y' = y + 3$, $\quad y(0) = 1$ 16. $(x^2 + 2)y' = y^2$, $\quad y(0) = 1$

In exercises 17-20, repeat the given exercise with $h=.05$.

17. exercise 9   18. exercise 10
19. exercise 11   20. exercise 12

In exercises 21-24, use the exponential decay formula $Q(t) = Q_0 e^{kt}$.

21. A radioactive substance has decay rate $k = -.2\,\text{hour}^{-1}$. If 2 grams of the substance is present initially, when will only half the original amount remain? NOTE: this is called the *half-life* of the substance.

22. In exercise 21, when will one-fourth the original amount remain? When will one-eighth remain? Explain what is meant by the statement "the half-life is independent of the initial amount."

23. If there are initially 2 grams of a substance present and the half-life is 1 hour, how much of the substance is left after 1 hour? 2 hours? 3 hours? 4 hours?

24. A common form of fossil dating is based on radioactive decay. Carbon-14 is found in living organisms and decays exponentially after death. The half-life is 5568 years. If a fossil is found to have 10% of its carbon-14 remaining, how old is the fossil? HINT: assume that there was 1 gram present at the time of death ($t=0$).

25. Write an automated program to do Euler's method. HINT: given h, determine how many steps are needed and loop (see your calculator manual) through Euler's method that many times.

26. In this exercise, we will use an improvement of Euler's method called (cleverly enough) the *improved Euler's method*. Using the same format as Euler's method, we will compute $y_{n+1} = y_n + h f_n$ where f_n is the slope of a line which approximates the actual slope of the solution. Instead of using $f_n = f(x_n, y_n)$ as in Euler's method this time we will have f_n be the average of the slopes $f(x_n, y_n)$ and $f(x_{n+1}, y_{n+1})$. Why would you expect this to be more accurate than Euler's method? Explain why $f_n = .5[f(x_n, y_n) + f(x_{n+1}, y_{n+1})] \approx .5[f(x_n, y_n) + f(x_n + h, y_n + hf(x_n, y_n))]$. We then get the formula

$$y_{n+1} = y_n + \frac{f(x_n, y_n) + f(x_n + h, y_n + hf(x_n, y_n))}{2} h$$

Rework exercise 9 using the improved Euler's method, and compare both approximations to the exact solution $y = e^{x^2}$

EXPLORATORY EXERCISE

Introduction

Among all the detective movies made, nobody has ever filmed the following. A murder has been committed, the ace detective examines the scene of the crime, coolly whips out a calculator, punches a few buttons and announces the time of death. The TI-81/85 may not make it in Hollywood, but we will use it below to play detective. We will use Newton's Law of Cooling, which states that the temperature of an object changes at a rate proportional to the difference between the temperature of the object and its environment. If $T(t)$ is the temperature at time t, then $T'(t) = k[T(t) - E]$ where E is the temperature of the environment and k is a constant.

Problems

At 6:00 we discover a secret agent bound and murdered. Next to him is a martini which got shaken before he could stir it. Room temperature is $70°$, and the martini warms from $60°$ to $61°$ in the 2 minutes from 6:00 to 6:02. If the secret agent's martinis are always served at $40°$, what is the time of death?

There are two estimates to be made, both of which we can use EULER for. First set $T(0) = 60$ and (by trial-and-error) determine k such that $T(2) = 61$. That is, guess a value for k (should it be positive or negative? large or small?). Plugging k and E into the differential equation, use EULER to estimate $T(2)$. How does it compare to 61? Based on the calculated value of $T(2)$, adjust the value of k and try again. Continue this until $T(2)$ is close enough to 61. Then, set $T(0) = 40$ and determine t such that $T(t) = 60$.

The time of death is t minutes before 6:00.

Further Study

In a course on differential equations, you will learn to find exact solutions to this and other problems arising from basic physical principles. Most calculus books also include a chapter on differential equations.

CHAPTER
4

Applications of Differentiation

4.1 Rootfinding Methods

Most students remember how much time they spent in their high school algebra class answering the question: for what values of x is

$$f(x) = ax^2 + bx + c = 0$$

The values (called *zeros* or *roots*) are, of course, found by using the now familiar quadratic formula. But, what if f were not a quadratic polynomial? You might well hope that either the question is irrelevant or that no one will ever ask you to solve such a problem. Unfortunately, the question *is* relevant and such questions need to be answered wherever calculus is applied, from engineering and the physical sciences to economics and business applications.

Example 1. A Difficult Rootfinding Problem

In the vibration of an elastic string, under certain conditions, the natural frequencies (i.e., one of the physical characteristics used to describe the vibration) are solutions, x, of the equation

$$f(x) = \tan(x) - x = 0$$

Notice that this is not an algebraic problem. That is, we cannot use the usual rules of algebra to solve for the value(s) of x. However, you can see from a simple graph (we'll do this later) that there are indeed values of x which satisfy the equation (in fact, there are infinitely many of them). But how can we find these values? In short, we can't, at least not exactly. The best that we can do in this case is to find approximate solutions. ∎

So, what we have is a compelling problem with no apparent means of solution. In this section, we will discuss several simple methods which can be employed with a programmable calculator and show how to use the graphics and programming features of the TI-81/85 to implement these methods effectively.

Recall that we had briefly discussed this problem in Chapter 1, when we showed how to use the TRACE function to locate (approximately) the x-intercepts of graphs. We also discussed how to use the TI-85 Solver to solve for roots. Well, what else is there to discuss then? There's no denying that we can find roots approximately by using a graph and the TRACE function, but with the need for repeated zooming, this is tedious and the results are crude, at best. If you have a TI-85, the built-in Solver is quite capable. It can be used to quickly and accurately solve a wide variety of rootfinding problems. But that is not enough. The Solver is a *black box*, whose inner workings are unknown to the user. While this may be acceptable if our only interest is to find answers to problems, the entire purpose of our discussion here is to find *understanding*. Further, the Solver does *not* always work and the advisory message is not always to be trusted.

Example 2. A Faulty Answer from the TI-85 Solver

If you have a TI-85, use the Solver to try to find a root of $f(x) = x^4 - 5x^3 + 9x^2 - 7x + 2$. First, draw a graph to see that there seems to be a root near $x = 0.9$. So, we use the Solver with the initial guess 0.9. Press $\boxed{\text{SOLVER}}$. Enter the equation

$$A = x{\wedge}4 - 5\ x{\wedge}3 + 9\ x{\wedge}2 - 7\ x + 2$$

on the line next to "eqn:" press $\boxed{\text{ENTER}}$ and a list of the variables will appear. Give A the value 0, x the value 0.9 and with the cursor still blinking on the x line, press $\boxed{\text{SOLVE}}$. The Solver returns the approximate root 1.0000435314113, with the advisory message "lft-rt=2E−13." This indicates that the Solver thinks that it has found an approximate root, where the left and right sides of the equation differ by only 2×10^{-13}. This would seem to be an acceptable answer and is not particularly remarkable until we observe that $f(x)$ factors:

$$f(x) = x^4 - 5x^3 + 9x^2 - 7x + 2 = (x-1)^3(x-2)$$

Thus, the only two roots are exactly $x = 1$ and $x = 2$. While $x = 1.0000435314113$ may not seem a poor approximation of 1.0 if you are using this to aim the throw of a baseball from center field to home plate, this is not particularly precise if you are instead aiming a spacecraft at the moon. This stands in sharp contrast to the confident advisory message "lft-rt=2E−13" displayed by the Solver. The moral of the story, of course, is that you should develop a healthy amount of skepticism for the results of the Solver (for finding roots or extrema, at least) or for the results of any other "black box" method, where the technique and the intermediate calculations are unknown.

∎

In this chapter, we will examine how roots are approximated. We'll discuss the various strengths and weaknesses of the methods we develop, and we'll look at problems for which the Solver as well as our other methods can get fooled. Further, we'll see how to follow the output of a rootfinding method to see when it is going astray.

THE METHOD OF BISECTIONS

Recall that the graph of a continuous function can be drawn without lifting the pencil from the paper. For such a function, if the graph is to pass from one side of the x-axis to the other, it must pass through the x-axis somewhere. The following result (a consequence of the Intermediate Value Theorem) is a more formal statement of this principle and should be fairly evident.

Theorem 4.1 Suppose that f is a continuous function on the interval $[a, b]$, and that $f(a) \cdot f(b) < 0$. [i.e., $f(a)$ and $f(b)$ have opposite signs]. Then, for some number c in (a, b), $f(c) = 0$.

This simple result is the basis for the most elementary method of numerical rootfinding. Given that f has opposite signs at a and b, we might guess that a root could be halfway in between a and b, i.e., at

$$c = \frac{1}{2}(a + b)$$

If not, then there must be a root in at least one of the intervals (a, c) or (c, b). To check if a root might be in (a, c), compare $f(a)$ and $f(c)$ for a change of sign. If there's no change, then there must be a root in (c, b). (Note that we say "a" root, as opposed to "the" root, since there may well be more than one root in the interval.) We then proceed to look at the value of f at the midpoint of the new interval and so on. The following algorithm, called the *method of Bisections*, is thus generated.

Step	Explanation
1. Check that $f(a) \cdot f(b) < 0$.	Look for a sign change to make certain that there's a root in (a, b).
2. Let $c = \frac{1}{2}(a + b)$.	Find the midpoint of $[a, b]$.
3. If $f(c) = 0$, stop.	Stop if you find a root.
4. If $f(a) \cdot f(c) < 0$ replace b by c and go to step 2.	Check $[a, c]$ for a sign change. If there is one, look for a root in the interval (a, c).
5. Otherwise, replace a by c and go back to step 2.	If there's no sign change in $[a, c]$ then there's one in $[c, b]$. Look for a root there.

Of course, in practice, the stopping condition, $f(c) = 0$, is only rarely realized. Usually, c is considered to be an acceptable approximation to a root if $f(c)$ is "small enough" in absolute value (how small is small enough is determined by the need for accuracy in the particular problem). Step 3 in the algorithm is then replaced by:

$$3a. \text{ If } |f(c)| < TOL, \text{ stop}$$

where TOL is some measure of closeness to zero, e.g., TOL=.01 if .01 is a "small enough" function value. Alternatively, we may be satisfied in knowing that a root is in the interval (a, b), where $(b - a)$ is sufficiently small. In this latter case, we replace step 3 with the step

$$3b. \text{ If } (b - a) < TOL, \text{ stop}$$

where, again, TOL is an acceptable tolerance.

We shall first illustrate the Bisections algorithm with an example done by hand. Work through Example 3 to make sure that you understand the progression from step to step. We will then proceed to TI-81/85 programs for Bisections.

Example 3. The Method of Bisections - Step by Step

Find a root of $f(x) = x^5 - 5x^3 + 3$ in the interval $[0,1]$. First note that $f(0) = 3$ and $f(1) = -1$. Since f is continuous (*All* polynomials are continuous!) and f has opposite signs at $a = 0$ and $b = 1$, then Theorem 4.1 implies that there must be a root in the interval $(0,1)$. For this particular problem, we'll agree to accept a solution accurate to 2 decimal places (i.e., we'll stop computing new values when $|b - a| < .01$).

Set $c = \frac{1}{2}(a + b) = .5$; $f(.5) = 2.4 > 0$. Since $f(1) < 0$, this means that there's a root between $a = .5$ and $b = 1$; we have $b - a = .5$.

Set $c = \frac{1}{2}(.5 + 1) = .75$; $f(.75) = 1.1 > 0$. (A root is between $a = .75$ and $b = 1$; $b - a = .25$.)

Set $c = \frac{1}{2}(.75 + 1) = .875$; $f(.875) = .16 > 0$. (A root is between $a = .875$ and $b = 1$; $b - a = .125$.)

Set $c = \frac{1}{2}(.875 + 1) = .9375$; $f(.9375) = -.4 < 0$. (A root is between $a = .875$ and $b = .9375$; $b - a = .0625$.)

Set $c = \frac{1}{2}(.875 + .9375) = .90625$; $f(.90625) = -.11 < 0$. (A root is between $a = .875$ and $b = .90625$; $b - a = .03125$.)

Set $c = \frac{1}{2}(.875 + .90625) = .890625$; $f(.890625) = .03 > 0$. (A root is between $a = .890625$ and $b = .90625$.)

Note that although $f(.890625)$ is not 0, there must be a root in the interval $(.890625, .90625)$. Thus, if we only need to have 2-digit accuracy, we can declare that the midpoint of that interval, $c = \frac{1}{2}(.890625 + .90625) = .8984375$, is an approximate root, since no number in the interval $(.890625, .90625)$ can be more than $b - c = .90625 - .8984375 \approx .008$ away from the center, c. (Why is that?) ∎

If greater accuracy is desired, then one must simply continue this process further. You will no doubt agree that this is a tedious procedure, at least when performed by hand. But, who needs to do computations by hand, or even manually, when we have a powerful machine like the TI-81/85 at our disposal? We suggest the following TI-81 program to perform the Bisections algorithm. Recall that the function name Y_1 must be taken from the Y-VARS menu, and that $\boxed{\leq}$ is found in the Test menu.

:BISECT   :.5*(A+B)$\boxed{\rightarrow}$ C   :A$\boxed{\rightarrow}$ X   :$\boxed{Y_1}$ $\boxed{\rightarrow}$ D
:C$\boxed{\rightarrow}$ X   :$\boxed{If}$ (D*$\boxed{Y_1}$)$\leq$0   :$\boxed{Goto}$ A   :C$\boxed{\rightarrow}$ A
:$\boxed{Goto}$ B   :$\boxed{Lbl}$ A   :C$\boxed{\rightarrow}$ B   :$\boxed{Lbl}$ B   :$\boxed{Disp}$ C

Program Step	Explanation
:BISECT	Name the program.
:.5*(A+B)$\boxed{\rightarrow}$ C	Store the midpoint of A and B in C.
:A$\boxed{\rightarrow}$ X	Store the contents of A in the variable X.
:$\boxed{Y_1}$ $\boxed{\rightarrow}$ D	Store the value of $f(x)$ in the variable D.
:C$\boxed{\rightarrow}$ X	Store the contents of C in the variable X.
:$\boxed{If}$ (D*$\boxed{Y_1}$)$\leq$0	Test for a sign change between A and C.
:$\boxed{Goto}$ A	Go to Label A if there is a sign change.
:C$\boxed{\rightarrow}$ A	If there is no sign change on [A,C] replace A by C.
:$\boxed{Goto}$ B	Go to Label B.
:$\boxed{Lbl}$ A	Label A.
:C$\boxed{\rightarrow}$ B	If there was a sign change on [A,C] replace B by C.
:$\boxed{Lbl}$ B	Label B.
:$\boxed{Disp}$ C	Display the value of the midpoint.

The corresponding TI-85 program is nearly identical. The contents of the program follow. When prompted for a name, enter BISECT. As usual, pay close attention to the x's and y's to make certain that you use lower cases. We suggest that you type in the function name y1 directly instead of taking it from the VARS menu. Also, recall that $\boxed{<}$ is found in the Test menu.

:.5*(A+B)$\boxed{\rightarrow}$ C   :A$\boxed{\rightarrow}$ x   :y1 $\boxed{\rightarrow}$ D   :C$\boxed{\rightarrow}$ x   :$\boxed{If}$ (D*y1)$\leq$0
:$\boxed{Goto}$ A   :C$\boxed{\rightarrow}$ A   :$\boxed{Goto}$ B   :$\boxed{Lbl}$ A   :C$\boxed{\rightarrow}$ B   :$\boxed{Lbl}$ B   :C

To run the above program (on either calculator), you must first store appropriate values in the variables A and B. (Recall that the method requires that A<B and $f(A) \cdot f(B) < 0$.) Also, you must enter the function f into the "Y=" list as Y1. Executing BISECT will then return the value of C to the Home screen. Continually updated values of the endpoints A and B are stored in those respective variables. Pressing the $\boxed{ENTER}$ key repeatedly will produce a screen full of successive approximations to the root.

At this point, you should run your BISECT program for the problem done by hand in Example 3. First, enter the function in the "Y=" list:

$$Y1 = x \wedge 5 - 5 * x \wedge 3 + 3$$

Next, store the endpoints of an interval containing a root in the variables A and B. For the present problem, such an interval is [0,1]. Press 0 $\boxed{\rightarrow}$ A $\boxed{ENTER}$ and 1 $\boxed{\rightarrow}$ B $\boxed{ENTER}$.

Execute $\boxed{BISECT}$ once; the value .5 is returned to the home screen. Continuing to execute BISECT (simply press $\boxed{ENTER}$ repeatedly) will produce smaller and smaller intervals containing a root. Repeating the program 6 more times yields the interval [.890625,.8984375]. The width of this interval is .0078125. Thus, if we are looking for an approximation to a root valid to 2 decimal places, the midpoint of the interval, .89453125, will suffice. If more accuracy is needed, we can continue to execute BISECT until we obtain an interval whose width is sufficiently small. (You can press B−A $\boxed{ENTER}$ to check. If the interval is not small enough, note that you can no longer continue with execution of the program by pressing $\boxed{ENTER}$. You will need to access BISECT from the PRGM menu as if starting from scratch.)

The midpoint of any sufficiently narrow interval will serve as an approximate root. [Enter (A+B)/2.] Actually, an interval whose width is twice the tolerance will do. (Why?)

If you continue executing BISECT for the above problem, you should arrive at the approximate root .8938274621, valid to the limits of the machine's accuracy. (Note that the TI-85 will do slightly better, but only at the cost of performing a few more steps.) Pay particular attention to the values of C displayed on the home screen. Watching these values, you can see how the method of Bisections homes in on a root. If you've been counting, it should have taken 36 repeated applications of BISECT (and perhaps a minute of your time) to arrive at this value. This does not seem too bad. After all, we are obtaining an approximate root, accurate to about 10 decimal places. Nonetheless, this is one of the most significant drawbacks to using Bisections in practice. Relative to other methods (several of which we will discuss later in this chapter), it is very slow. Of further concern is that Bisections can be used only when we can find values A and B between which f changes sign. Unfortunately, this cannot always be done.

Example 4. A Problem to Which Bisections Cannot Be Applied

Consider $f(x) = x^2$. Note that $f(x) > 0$, except for $x = 0$, and, hence, we cannot use Bisections to locate the root $(x = 0)$.   ∎

NEWTON'S METHOD

The discerning reader will undoubtedly wonder why another rootfinding method is necessary. First, as noted above, Bisections is a relatively slow method. We'll see that we can use the power of calculus (in particular, the derivative) to develop faster and more accurate methods. More generally, good problem solvers need an assortment of mathematical tools, just as all craftsmen need a variety of tools to perform their jobs.

A generally faster method for approximating roots of a function, called *Newton's method*, works in the following way. Using some graphical or numerical evidence, we first make a guess as to where a root is located. Specifically, suppose that a given function f has a root located near some value x_0. Assuming that f is differentiable at x_0, we can draw in the tangent line to $y = f(x)$ at $x = x_0$ (see Figure 4.1).

The slope of the tangent line is $f'(x_0)$ and hence its equation is

$$m_{\text{tan}} = f'(x_0) = \frac{y - f(x_0)}{x - x_0}$$

Next, follow the tangent line to the point where it intersects the x-axis. In the figure, this appears to be closer to the root than x_0. Call the x-coordinate of this point x_1. Since this is the point of intersection with the x-axis, this corresponds to

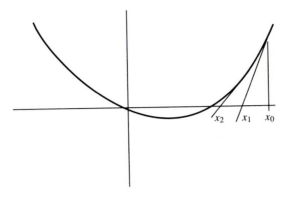

FIGURE 4.1

the point on the line where $y = 0$. From the above equation for $f'(x_0)$, we get

$$x_1 = x_0 - \frac{f(x_0)}{f'(x_0)}$$

Rather than being satisfied with this modest improvement over our original guess, we repeat the above procedure, replacing x_0 with x_1, to obtain a new (and, ideally, further improved) approximation x_2. Continue this process until no further progress is made. This generates a sequence of approximations,

$$x_{n+1} = x_n - \frac{f(x_n)}{f'(x_n)} \qquad n = 0, 1, 2, \ldots$$

This is the general form of Newton's method. Let us return to Example 3 and compare the performance of Bisections with this new method.

Example 5. Newton's Method

Find a root of $f(x) = x^5 - 5x^3 + 3$, using Newton's method with an initial guess of $x_0 = 1$. First, notice that $f'(x) = 5x^4 - 15x^2$. Newton's method becomes

$$x_{n+1} = x_n - \frac{x_n^5 - 5x_n^3 + 3}{5x_n^4 - 15x_n^2}$$

Using $x_0 = 1$, we obtain the following table. Be sure to check these results yourself. You can do this easily by entering the function $x - (x \wedge 5 - 5x \wedge 3 + 3)/(5x \wedge 4 - 15x \wedge 2)$ into the "Y=" list as Y1 and by using the program FEVAL discussed in Chapter 1.

n	x_n	$f(x_n)$
1	.9	$-.05451$
2	.8938542195	$-.000235268728$
3	.8938274626	$-.000000004486$
4	.8938274621	0
5	.8938274621	0

Notice that the last two values are identical, at least in the 10 decimal places displayed (about the most precision one can expect from the TI-81). Since the value of the function is also reported as 0, you should begin to suspect that you have obtained a good approximation to a root. You should also notice that Newton's method produced an approximation in many fewer steps and with far less effort than the method of Bisections. Finally, we should observe that a good time to stop computing new approximations would seem to be when two successive approximations are sufficiently close (i.e., when $|x_{n+1} - x_n| < \text{TOL}$ for some acceptable tolerance TOL), as is the case for x_4 and x_5 in the table.   ∎

CAUTION: In practice, even a reported function value close to zero does not guarantee that the approximation is close to an actual root. We will examine this further in the exercises and in section 4.2.

We now give a TI-81 program to perform Newton's method.

:$\boxed{\text{Ans}}$ $\boxed{\rightarrow}$ X   :X $-$ $\boxed{Y_1}$ / $\boxed{Y_2}$:$\boxed{\text{Disp}}$ $\boxed{\text{Ans}}$

Program Step	Explanation
:NEWTON	Name the program.
:$\boxed{\text{Ans}}$ $\boxed{\rightarrow}$ X	Store the last value entered in the variable X.
:X $-$ $\boxed{Y_1}$ / $\boxed{Y_2}$	Compute the Newton step.
:$\boxed{\text{Disp}}$ $\boxed{\text{Ans}}$	Display the value on the home screen.

The TI-85 program is nearly identical. When prompted for a name, enter NEWTON. The contents of the program follow:

$$: \boxed{\texttt{Ans}} \boxed{\rightarrow} \text{x} \quad : \text{x} - \text{y}1/\text{y}2$$

Before running the above program, you will need to store expressions corresponding to $f(x)$ and $f'(x)$ in the functions Y1 and Y2, respectively, in the "Y=" list. Then, enter an initial guess on the home screen (simply type in the value and press $\boxed{\texttt{ENTER}}$). Each time $\boxed{\texttt{NEWTON}}$ is executed, the next Newton step (x_{n+1}) is computed and displayed on the home screen.

PROGRAM NOTE: Although the TI-81/85 has the facility to compute approximate (numerical) derivatives (recall the discussion in section 3.2), it is inefficient to have the machine do so in each step of a Newton's method procedure. Therefore, we have instead instructed the user to store the derivative as a separate function in the "Y=" list.

Now, key in the program and test it on the example given above. First, store $f(x)$ and $f'(x)$ in your "Y=" list, as Y1 and Y2, respectively. Your "Y=" list should look like the following:

$$Y1 = x \wedge 5 - 5 \, x \wedge 3 + 3$$
$$Y2 = 5 \, x \wedge 4 - 15 \, x \wedge 2$$

To run the program for the current example, press 1 $\boxed{\texttt{ENTER}}$ and execute the program NEWTON. The successive approximations are displayed on the home screen, each time the program is executed (after the first time, this is done by simply pressing $\boxed{\texttt{ENTER}}$). Continue until the last two displayed values are within the desired tolerance. The root is approximately .893827462. Pay particular attention to how the approximation improves at each step of the process and compare this with the behavior of the Bisections method.

CAUTION: Since Newton's method is not guaranteed to work, it's a good idea to further test the validity of the results by computing the value of the function f at the suspected approximate root. This is most easily accomplished by storing and executing the following simple TI-81 program.

$$: \text{FEVA2} \quad : \boxed{\texttt{Ans}} \boxed{\rightarrow} \text{X} \quad : \boxed{\texttt{Y}_1} \quad : \boxed{\texttt{Disp}} \boxed{\texttt{Ans}}$$

The corresponding TI-85 program follows (when prompted for a name, enter FEVA2).

$$: \boxed{\texttt{Ans}} \boxed{\rightarrow} \text{x} \quad : \text{y}1$$

Executing FEVA2 will compute the value of the function f at Ans (the last value computed and sent to the home screen).

Notice that even for a very small tolerance, say TOL = .00000000001, the Newton's method program for this problem takes only a few steps. Compare this with the number of steps required for the Bisections method for this problem. This is one of the main advantages of Newton's method. It will usually take many fewer steps than the method of Bisections to achieve the same accuracy. For complicated problems, there can be a substantial difference between the two methods. Thus, Newton's method is typically favored over the method of Bisections. In automated programs (i.e., ones which will run automatically until a specified tolerance is reached), Bisections programs typically take much longer to run than programs for Newton's method. (Can you think of a time when Bisections would be the method of choice? Hint: Newton's method requires one to compute a derivative.)

Now that we've praised the benefits of Newton's method, we want to be perfectly honest with you. Newton's method does have a negative side: it does not always work. In practice, the initial guess must be chosen fairly close to a root in order that the successive approximations will home in on a root. Just *how* close to the root it must be chosen, though, varies from problem to problem. The answer to this question would be of little use anyway, since, in practice, we do not know where the root is. (If we did, then we wouldn't be employing an approximation method like Newton's method!)

Example 6. An Initial Guess that Does Not Work

Consider the function $f(x) = x^3 - 3x^2 + x - 1$. Notice that there is a root somewhere in the interval (1,3). (Why is that?) Using Newton's method with the initial guess $x_0 = 1$, we get $x_1 = 0$ and $x_2 = 1$ and so on. The values alternate back and forth between 0 and 1, neither one of which is a root. Try the example for yourself. If we instead start with a slightly better initial guess, say $x_0 = 2$, Newton's method will converge quickly to the value 2.769292354.

∎

Although Newton's method is generally an accurate and efficient rootfinding method, you should be aware that there are other bad things that can happen with Newton's method. In Example 7, the successive values will wander away from the only root and tend toward minus infinity.

Example 7. Newton's Method Wanders Away from the Root

Consider the function

$$f(x) = \frac{(x-1)^2}{x^2+1}$$

Obviously, f has only one root, at $x = 1$. Using our Newton's method program with $x_0 = -2$, we get the results (the x_n values have been rounded off):

n	x_n
1	-9.5
2	-65.9
3	-2302
4	-2654301
5	-3.5 E12
6	-3.1 E24

Notice that the last two values are very large in absolute value and are getting rather close to the outer reaches of the accuracy of the TI-81/85 (in fact, the TI-81 is at its limit, while the TI-85 returns the value $x_6 = -6.2$ E24). You should learn quickly to be skeptical of any reported approximate root that is so large in absolute value.

If we use the improved initial guess $x_0 = -1$, the program immediately returns an error, caused by an attempted division by 0. (Why?) Finally, with the even better initial guess $x_0 = 0$, Newton's method will converge to the root, but uncharacteristically slowly. (Try this. We'll look further into this type of behavior later.) ∎

On the whole, Newton's method should be viewed as a very useful and accurate method for finding approximate roots, when used with a bit of caution. In the exercises, we will demonstrate more cases where the method fails to yield an acceptable answer, as well as a number of typical examples where things work just fine.

USE OF GRAPHICS FOR DETERMINING INITIAL GUESSES

There is one important question which we have so far avoided. How do you come up with the initial guess(es) needed for either the method of Bisections or for

Newton's method? One suggestion might be to randomly guess. However, we have already seen that the method of Bisections requires good input to obtain accurate answers in a reasonable length of time and that Newton's method may require a good initial guess just to work at all. We therefore require an approach to finding initial guesses which is more sensible than random guessing. The graphics capabilities of the TI-81/85 are well suited for this purpose.

Example 8. Using Graphics to Find Initial Guesses

Find the roots of $f(x) = x^4 + 5x^3 - 5x^2 + 7x - 6$. To get an idea of where the root(s) may be located, we draw a graph. First, enter the equation to be plotted into the "Y=" list (you'll need to do this anyway in order to use your rootfinding methods):

$$Y1 = X \wedge 4 + 5 X \wedge 3 - 5 X \wedge 2 + 7 X - 6$$

To insure that you have reset the graphics window to its default setting, press ZOOM 6 (Standard) on the TI-81 or ZOOM ZSTD on the TI-85. See Figure 4.2 for the TI-81 graph. Although only part of the graph fits on the screen, you can clearly see two roots. Moving the cursor along the x-axis, there appears to be a root at (exactly) $x = -6$ and another root between $x = .7368$ and $x = .9474$.

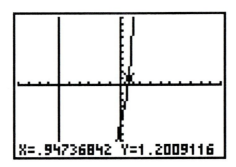

X=.94736842   Y=1.2009116

FIGURE 4.2

Note that even though the graph makes it look as if there is a root at exactly $x = -6$, we should not be too quick to accept this as even an approximate root: observe that $f(-6) = -12$! However, if we use this as an initial guess for Newton's method, (i.e., let $x_0 = -6$), we quickly obtain the approximate root $x = -6.045702144$, where $f(x) = -7.3E-10$. Of course, we could also use our BISECT program, although this is much slower. If you have a TI-85, recall that you may use the ROOT command (see Example 1 of section 1.3).

Finally, to locate the root between $x = .7368$ and $x = .9474$, we choose $x_0 = 0.8$ as our initial guess in Newton's method. We quickly obtain the approximate root $x = 0.8558884183$, where $f(x) = -1.0E-12$. The fact that Newton's method converged to these two values, combined with the fact that these values correspond to what we expected from the graph, gives us confidence that we have indeed found approximate roots.   ∎

Example 9. A Problem with Using Graphics to Find the Initial Guess

Consider $f(x) = x^3 - 11x^2 + x - 5$. In the initial TI-81 graph (see Figure 4.3a), no roots are apparent. In fact, very little of the graph is displayed on the screen. Yet, we know that f must have at least one (real) root, since it is a cubic polynomial. One remedy is to zoom out some. On the TI-81, press ZOOM 3 (Zoom Out) and (when the graph is displayed) press ENTER . (On the TI-85, press ZOOM ZOUT ENTER .) The corresponding TI-81 graph is displayed in Figure 4.3b (where the zoom factors are Xfact=Yfact=2). A root is now clearly visible near $x = 10$.

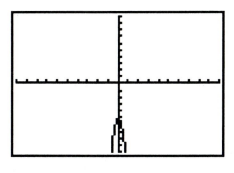

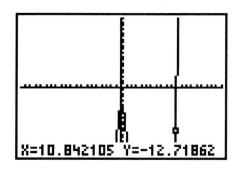

X=10.842105 Y=-12.71862

FIGURE 4.3a   FIGURE 4.3b

To get a better handle on the behavior of that part of the function whose graph appears on the screen, we zoom in (press ZOOM 1 (Box) on the TI-81 or ZOOM BOX on the TI-85 and use the arrow keys to get the zoom box shown in Figure 4.4a). The zoomed graph is shown in Figure 4.4b.

NOTE: Although the graph would seem to indicate that a root is located near $x = 10.85$, we should not accept this as an approximate root without further evidence. [Simply compute $f(10.85)$ to see why this is unacceptable.] However, if we use this as the initial guess for Newton's method, we get (after several steps) the approximate root $x = 10.95037662$, where $f(x) = 0$.

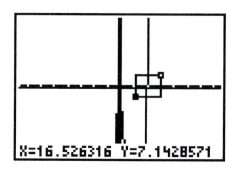

X=16.526316 Y=7.1428571

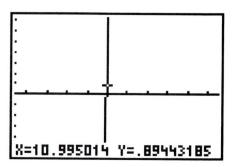

X=10.995014 Y=.89443185

FIGURE 4.4a   FIGURE 4.4b

Having found one root, it still remains for you to see if there are any other roots. We will turn our attention to this in section 4.2. For the moment, we shall be content with locating a single root.   ∎

THE SECANT METHOD

We have seen that Newton's method can be a very useful tool for approximating roots of functions. We have also pointed out several significant limitations. First, the initial guess, x_0, must be chosen sufficiently close to the root (and we never know when a given guess is *sufficiently* close). Second, the method requires us to compute a derivative. The latter requirement can be quite restrictive: the function may not be differentiable or the derivative computation may be prohibitively complicated. If this is the case, we could always use Bisections. However, Bisections is exceedingly slow and can be used only if we can find numbers A and B for which $f(A)$ and $f(B)$ have opposite signs. We now present a method which has most of the advantages of Newton's method, but does not require us to compute a derivative.

Given two initial guesses, x_0 and x_1 (not necessarily bracketing a root), draw the secant line joining the two points. The slope of this line is

$$m_{sec} = \frac{f(x_1) - f(x_0)}{x_1 - x_0}$$

The equation of this secant line is then

$$\frac{y - f(x_1)}{x - x_1} = \frac{f(x_1) - f(x_0)}{x_1 - x_0}$$

As with Newton's method, we follow the line to where it crosses the x-axis (i.e., where $y = 0$; see Figure 4.5). Call the x-coordinate of this new point x_2. We get

$$x_2 = x_1 - f(x_1)\frac{x_1 - x_0}{f(x_1) - f(x_0)}.$$

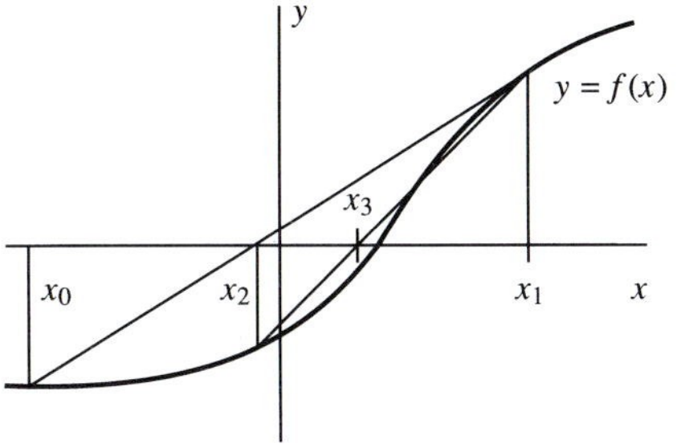

$\rho.\,u$

FIGURE 4.5

In Figure 4.5, x_2 appears to be closer to the root than either x_0 or x_1. We can repeat the procedure over and over, each time using the latest two values to compute a new approximation. We get:

$$x_{n+2} = x_{n+1} - f(x_{n+1})\frac{x_{n+1} - x_n}{f(x_{n+1}) - f(x_n)} \qquad n = 0, 1, 2, \ldots$$

This is known as the *Secant method*. Note the similarity with Newton's method. In this case, we start with two initial guesses and then approximate the slope of the tangent line at x_{n+1} by the slope of the secant line joining the points corresponding to x_n and x_{n+1},

$$f'(x_{n+1}) \approx \frac{f(x_{n+1}) - f(x_n)}{x_{n+1} - x_n}$$

In practice, the Secant method will converge nearly as fast as Newton's method. Its main advantage over Newton's method is that it does not require the computation of a derivative. For some problems, this is a decisive advantage. We now return to an earlier example, to compare our new method with Bisections and with Newton's method.

Example 10. Secant Method

Find a root of $f(x) = x^5 - 5x^3 + 3$, using the Secant method with initial guesses $x_0 = 1$ and $x_1 = 0$. Using the TI-81/85, we obtain the results in the table below (be certain to check these yourself). We ceased computing new approximations when we ran across an x-value for which the reported function value was zero. Notice that the method takes a few more steps than Newton's method (8 steps compared to the

4 or 5 steps of Newton's method or the 36 steps of Bisections required to obtain the same accuracy). This somewhat slower convergence is typical of the Secant method.

n	x_n	$f(x_n)$
2	0.75	1.13
3	1.202	−3.17
4	0.8685	0.2185
5	0.88999	0.0336
6	0.89389907	−0.00063
7	0.8938272655	0.0000017
8	0.8938274621	0.000000000088
9	0.8938274621	0.0

We suggest the following TI-81 program for the Secant method.

:SECANT   :A $\boxed{\rightarrow}$ X   :$\boxed{Y_1}$ $\boxed{\rightarrow}$ Y   :B $\boxed{\rightarrow}$ X

:X − $\boxed{Y_1}$ *(B−A)/($\boxed{Y_1}$ −Y) $\boxed{\rightarrow}$ T   :B $\boxed{\rightarrow}$ A   :T $\boxed{\rightarrow}$ B   :$\boxed{Disp}$ B

Program Step	Explanation
:SECANT	Name the program.
:A $\boxed{\rightarrow}$ X	Store the contents of A into X.
:$\boxed{Y_1}$ $\boxed{\rightarrow}$ Y	Compute $f(x)$ and store this in Y.
:B $\boxed{\rightarrow}$ X	Store the contents of B into X.
:X − $\boxed{Y_1}$ *(B−A) /($\boxed{Y_1}$ −Y) $\boxed{\rightarrow}$ T	Compute the secant step and store the value in the variable T.
:B $\boxed{\rightarrow}$ A	Replace A with B.
:T $\boxed{\rightarrow}$ B	Replace B with the new value T.
:$\boxed{Disp}$ B	Display the new secant step.

The corresponding program for the TI-85 follows. (We suggest that you name the program SECANT.)

:A $\boxed{\rightarrow}$ x   :y1 $\boxed{\rightarrow}$ Y   :B $\boxed{\rightarrow}$ x   :x−y1 *(B−A)/(y1−Y) $\boxed{\rightarrow}$ T
:B $\boxed{\rightarrow}$ A   :T $\boxed{\rightarrow}$ B

To run the above programs, you first need to store the initial guesses x_0 and x_1 in the variables A and B, respectively. You must also store the function f in the "Y=" list as Y1. Each time that the program is run, the next secant step, x_{n+1}, is computed and returned to the home screen. As with Newton's method, you should continue to compute new approximations until two successive values are within some prescribed tolerance.

As with Newton's method, the Secant method will usually require only a few steps, even for a very small value of the tolerance. Take a few minutes now to key in the program and to test it out by computing the values in the last table. Continue to execute the program until two successive values displayed on the home screen are identical. Be sure to pay particular attention to how the approximation is improving at each step and compare this with the behavior of Newton's method and the method of Bisections.

The only advantage of the Secant method over Newton's method is that it does not require us to compute a derivative. Both methods may fail to work for a given problem. Much like Newton's method, the Secant method requires initial guesses that are sufficiently close to a root in order to guarantee convergence to that root. In practice, we can use the graphical techniques described earlier to arrive at these guesses.

We have now presented three different methods for approximating roots and given TI-81/85 programs for each one. Each has advantages and disadvantages and we have pointed these out, where possible. With the three methods given and the hints presented for finding an initial guess(es) using the TI-81/85's graphing routines, you are now armed with all of the tools necessary for locating roots, with one exception. We have only discussed how to find *a* root of a function − not *all* the roots of a particular function. We will examine this question in the next section. By working carefully through the exercises, you will gain an appreciation for the various methods presented, as well as learn some of the shortcomings of each one. In this way, you will be prepared to deal effectively with a wide variety of rootfinding problems.

Exercises 4.1

In exercises 1-4, use Bisections, Newton's method and the Secant method to find approximate roots of the given function in the indicated interval. Use a tolerance of .0001. Compare the rates of convergence, as measured by the number of iterations required to achieve the desired tolerance.

1. $x^3 - 4x^2 - 8x - 2$, $[-2, -1]$ 2. $x^3 + 2x^2 - 49x - 8$, $[-1, 0]$
3. $-x^6 + 4x^4 - 2x^3 + 8x + 2$, $[2,3]$ 4. $x^4 - 7x^3 - 15x^2 - 10x - 1410$, $[10,11]$

In exercises 5-8, rework the indicated exercise by finding a root *outside* the given interval.

5. exercise 1   6. exercise 2

7. exercise 3   8. exercise 4

In exercises 9-10, use a rootfinding method to approximate the given root.

9. $\sqrt{3}$ (solve $x^2 - 3 = 0$)   10. $\sqrt[3]{2}$ (solve $x^3 - 2 = 0$)

In exercises 11-14, rewrite the given equation in the form $f(x) = 0$ and use a rootfinding method to approximate a solution in the interval.

11. $\sqrt{x^2 + 1} = x^3 - 3x - 1$, $[2,3]$ 12. $x^2 - 5 = \dfrac{\sqrt{x^2 + 1}}{x + 7}$, $[2,3]$

13. $\cos x = x$, $[0, \pi/2]$ 14. $\sin x = x^2 - 1$, $[-\pi/2, 0]$

In exercises 15-19, the indicated method fails in spite of the fact that there is a root in the indicated interval. Explain why the method fails, and explain how the root can be found.

15. $4x^3 - 7x^2 + 1$, $[0,1]$,   Newton's method with $x_0 = 1$.

16. $4x^3 + 12x^2 - 15x + 4 = 0$, $[0,1]$,   Bisections with $a = 0$ and $b = 1$.

17. $x^3 + 4x^2 - 19x + 15 = 0$, $[1,2]$,   Secant method with $a = 1$ and $b = 2$.

18. $\dfrac{4x^2 - 8x + 1}{4x^2 - 3x - 7} = 0$, $[1,2]$,   Newton's method with $x_0 = 1.5$.

19. $\dfrac{x^3 + x^2 - 2x - 2}{x^2 + 3x - 4} = 0$, $[0,2]$,   Bisections with $a = 0$ and $b = 2$.

20. For $f(x) = 2x - \dfrac{400}{x}$, we have $f(-1) \cdot f(2) < 0$. What happens if we attempt to run Bisections with $a = -1$ and $b = 2$? Is there actually a root in $[-1, 2]$?

In exercises 21-24, show that $x = 1$ is the only root and compare the rates of convergence of Newton's method with $x_0 = 0$.

21. $x^3 - x^2 + 4x - 4 = 0$

22. $x^4 - 2x^3 + 2x^2 - 2x + 1 = 0$

23. $x^4 - 3x^3 + 4x^2 - 3x + 1 = 0$

24. $x^3 - 2x^2 + 2x - 1 = 0$

25. When light passes from one medium to another, it refracts according to Snell's Law $\dfrac{v_1}{v_2} = \dfrac{\sin \theta_1}{\sin \theta_2}$ where v_i is the velocity of light in the ith medium and θ_i is the angle from the vertical. In the figure below, a person is looking at an underwater object. Using

$$\sin \theta_1 = \frac{x}{\sqrt{25 + x^2}} \qquad \sin \theta_2 = \frac{4 - x}{\sqrt{64 + (4 - x)^2}}$$

and $v_2 = .75v_1$, find x using the Secant method (why would this be simpler than Newton's method?). Also, find d, which is how far off the person's perception of the object is.

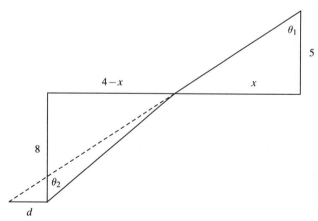

FIGURE FOR EXERCISE 25

26. A tennis serve hit from a height of 8 feet at an angle of θ below the horizontal will be successful (neglecting spin) if θ satisfies $t_1 < \cos \theta < t_2$ where

$$8t_1^2 - 39t_1 \sqrt{1 - t_1^2} = .95 \qquad 8t_2^2 - 60t_2 \sqrt{1 - t_2^2} = 2.25$$

Find θ_1 and θ_2 such that $\theta_1 < \theta < \theta_2$. For more details on tennis, see *Tennis Science for Tennis Players* by Howard Brody.

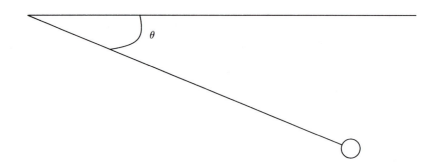

FIGURE FOR EXERCISE 26

27. To automate program BISECT, we only need to insert a program loop around the existing statements. In particular, insert $\boxed{\texttt{Lbl}}$ L at the top [before the line .5*(A+B)$\boxed{\rightarrow}$ C], and insert the following two statements between $\boxed{\texttt{Lbl}}$ B and $\boxed{\texttt{Disp}}$ C: $\boxed{\texttt{If}}$ B$-$A$\geq$T and $\boxed{\texttt{Goto}}$ L. Test the program on Examples 3 and 4. In what way(s) is this better or worse than the original BISECT?

28. Write a program to automate program NEWT, and test it on Examples 5-8.

29. Write a program to automate program SECANT. Test this on Example 10.

30. In program BISECT, our test was "$\boxed{\texttt{If}}$ (D*$\boxed{\texttt{Y}_1}$)$\leq$0." What would happen if "$\leq$" were replaced by "$<$" in this line? Try this substitution on Example 3 and on the following problem: $f(x) = x^2 - 1$, $a = 0$ and $b = 2$.

EXPLORATORY EXERCISE

Introduction

So far we have been finding one root at a time. In section 4.2, we will investigate how to find *all* of the roots. In this exercise, we will examine the behavior of Newton's method in a case where we know that there are three roots: $x(x - 1)(x - 2) = x^3 - 3x^2 + 2x = 0$. Clearly, the roots are 0, 1 and 2.

Problems

Try Newton's method with $x_0 = .1$, $x_0 = 1.1$ and $x_0 = 2.1$. All of these values are close to a root, and certainly nothing unusual happens. But, try $x_0 = .54$, $x_0 = .55$ and $x_0 = .56$. Are you surprised by the results? Examine the graph of $y = x^3 - 3x^2 + 2x$ and try to explain what happened.

If your curiosity has been piqued, then the next step is natural. We want to describe all starting values x_0 such that Newton's method converges to 0. This is called the *basin of attraction* of 0. Start by determining which root Newton's method converges to from $x_0 = 0, .01, .02, ..., .99, 1.0$. We suggest writing a general program for taking 100 steps between $x = A$ and $x = B$. (Such a program is given in the back of the book, but try writing one yourself!) Most of the basin boundaries are well defined, but there is some confusion between $x = .5$ and $x = .6$ (as we have already seen). It is reasonable to believe that all we need to do is magnify $[.5, .6]$. That is, run Newton's method with $x_0 = .5, .501, .502, ..., .599, .6$. Again, some clear boundaries emerge, but the picture is not sharp between .55 and .56. If you then magnify $[.55, .56]$, you will find erratic behavior in $[.552, .553]$. Continue this process and you will always see confusion in 1 out of 10 subintervals. Can we ever accurately determine the basin boundaries?

Further Study

We have seen a simple formula (Newton's method) produce very complicated behavior. This is a dominant characteristic of the exciting mathematical field of *chaos and fractals* (see Barnsley's *Fractals Everywhere*). The basins of attraction for complex roots are often quite beautiful. The picture below is generated from Newton's method for the root $z = 1$ of the complex variable equation $z^3 - 1 = 0$.

4.2 Multiple Roots

In section 4.1, we presented several different methods for approximating roots of a function. One question which we did not answer there was how to find *all* of the roots of a given function. The more basic question is to determine just *how many* roots a given function has. Unfortunately, the general theory surrounding this question is rather incomplete. We shall examine some examples and give some hints here. Another question which plagues the numerical approximation of roots is: what happens when a function has a root of multiplicity greater than 1 at a given point [e.g., $f(x) = (x - 2)^3$ has a root of multiplicity 3 at $x = 2$]? We will see that in this case, the speed and accuracy of both Newton's method and the Secant method will be reduced considerably, while Bisections may fail to work at all.

HOW MANY ROOTS ARE ENOUGH?

Obviously, many functions have more than one root. A reasonable question may be: how many roots are there? Unfortunately, there is no easy answer for this. Even for the familiar, relatively simple case of polynomial functions, the theory is inadequate. We do have:

Theorem 4.2 A polynomial p_n of degree n has at most n roots.

This says, for example, that a polynomial of degree 5 has at most 5 roots. Recalling that complex roots of a polynomial must come in conjugate pairs (i.e., if $a + b \cdot i$ is a root, then $a - b \cdot i$ is also a root), we see that polynomials of odd degree must have at least one real root. Hence, a polynomial of degree 5 could have 1,2,3,4 or 5 distinct real roots. For instance,

$$f(x) = (x - 1)(x - 2)(x - 3)(x - 4)(x - 5) \qquad \text{(5 roots)}$$
$$f(x) = (x - 1)^2(x - 2)(x - 3)(x - 4) \qquad \text{(4 roots)}$$
$$f(x) = (x - 1)^3(x - 2)(x - 3) \qquad \text{(3 roots)}$$
$$f(x) = (x - 1)^4(x - 2) \qquad \text{(2 roots)}$$
$$f(x) = (x - 1)^5 \qquad \text{(1 root)}$$

So, even for the familiar and relatively simple case of polynomials, we may not know how many roots there are without actually factoring the polynomial (which we can do in only a small number of cases, in practice). For more general functions, the answer is even less clear. However, we can use the graphics capabilities of the TI-81/85 to help answer the question.

Example 1. Using Graphics to Determine the Number of Roots

Consider $f(x) = \sin x - x^2 + 1$. If we draw the graph of $y = f(x)$, we can clearly see two roots (see Figure 4.6a for a TI-85 graph): one in the interval $(-1, 0)$, the other in the interval $(1,2)$. But, are there any roots that we don't see? That is, are there any roots outside of those displayed in the current graphics window? We can resolve this question easily for this particular example by plotting the graphs of $y = \sin x$ and $y = x^2 - 1$ simultaneously. Why? Well, notice that the equation

$$f(x) = \sin x - x^2 + 1 = 0$$

is equivalent to

$$\sin x = x^2 - 1$$

i.e., roots of f correspond to intersections of the two graphs $y = \sin x$ and $y = x^2 - 1$.

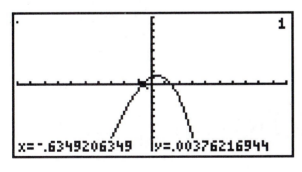

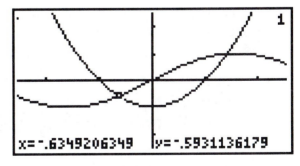

FIGURE 4.6a   FIGURE 4.6b

The TI-85 plot of the two superimposed graphs (Figure 4.6b) clearly shows the two points of intersection. Since we know that the graph of $y = x^2 - 1$ is a parabola opening up and since the graph of $y = \sin x$ oscillates back and forth between -1 and 1, we can easily infer from the plot that there are no other points of intersection. Of course, for this simple graph, we might have as easily drawn the graphs freehand. However, having done this using the TI-81/85, we have the added advantage that we can use the TRACE function to obtain initial guesses for the roots and then use these in one of our rootfinding schemes.

In the present case, the points of intersection appear to be near $(-.6, -.5)$ and $(1.4, 1)$. Using the x-coordinates of these points as initial guesses for Newton's method, we get the approximate roots:

$$x = -0.6367326508, \qquad \text{where } f(x) = 0.0$$

and

$$x = 1.409624004, \qquad \text{where } f(x) = -1.0 \times 10^{-12}$$

From the foregoing discussion, these are seen to be the only roots of f. ∎

We hasten to add that the preceding example, while not typical of all rootfinding problems, is of a type often encountered in applications. The suggestion that you rewrite $f(x) = 0$ as $g(x) = h(x)$ (for some appropriate selection of h and g) and draw superimposed graphs of g and h will help in a number of situations. This is particularly useful when the function f has both a periodic term (in the foregoing case, $\sin x$) and a term which is not oscillatory (in this case, $-x^2 + 1$).

A good test of your rootfinding skills is to solve the following problem.

Example 2. A Difficult Rootfinding Problem Solved

Find values of x for which $f(x) = \tan x - x = 0$. Certainly, there is no way of solving the problem algebraically, although it's clear that $x = 0$ is a root. Are there any others? The initial graph provided by the TI-81 (Figure 4.7a) gives us a somewhat confusing picture. The graph clearly passes through the origin (where we already know that there is a root). There are also some vertical or nearly vertical segments of the graph which cross the x-axis. But are these locations of roots, or do the line segments result from the TI-81 trying to connect two points on either side of a vertical asymptote? (Recall that $\tan x$ and hence f has vertical asymptotes at $x = \pi/2, 3\pi/2$, etc.) We get a slightly better graph by superimposing the graphs of $y = \tan x$ and $y = x$ (see Figure 4.7b for the TI-81 graph).

Still, the locations of the intersections (and hence the roots of f) are not entirely clear. At this stage, you might be led by the relative simplicity of the functions to draw the superimposed graphs of $y = x$ and $y = \tan x$ freehand (see Figure 4.8).

From Figure 4.8, we can see that there are infinitely many roots, since the tangent is periodic and blows up to infinity at $x = \pi/2, 3\pi/2, 5\pi/2,...$ For the sake of simplicity, we will concern ourselves here only with the case $x > 0$. As can be seen from the graph, the points of intersection are fairly close to the points $x = \pi/2$,

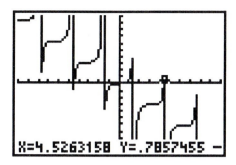

FIGURE 4.7a

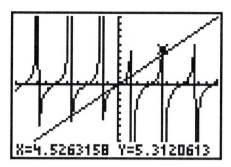

FIGURE 4.7b

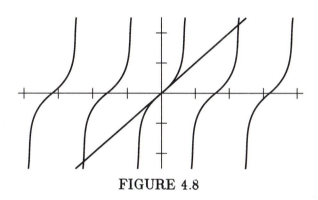

FIGURE 4.8

$3\pi/2$, $5\pi/2$,... In fact, the larger x gets, the closer the points of intersection (i.e., the roots of f) get to the points $x = \pi/2$, $3\pi/2$, $5\pi/2$, ... While it is of some significance to notice this, this does not help us find the roots. We cannot even use these values as initial guesses for any of our methods, since $\tan x$ is undefined at all of these points. However, we can still use the less revealing graph provided by the TI-81 to obtain acceptable initial guesses.

We return to the graph of $y = f(x) = \tan x - x$ (remember to "turn off" or clear out the function Y2 in the "Y=" list, so that only the graph of f is plotted). Use the ZOOM BOX function to zoom in on the apparent location of the first positive root (see Figures 4.9a and 4.9b for successive zooms). We can now see clearly that there is a root near $x = 4.5$. Using $x_0 = 4.5$ in our Newton's method program, we obtain the approximate root $x = 4.493409458$.

Before looking for the next positive root, reset the display parameters [press ZOOM 6 (Standard) on the TI-81 or press ZOOM ZSTD on the TI-85]. We again use the ZOOM BOX function to zoom in on the location of the next positive root (see Figures 4.10a and 4.10b). We now clearly see a root near $x = 7.73$. Using this as an

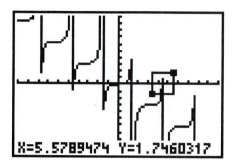

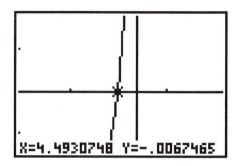

FIGURE 4.9a   FIGURE 4.9b

initial guess for Newton's method, we get the approximate root $x = 7.725251837$.

We have now found both of the positive roots shown in the initial graph (Figure 4.7a). The simplest way to look for roots farther to the right is to change the range. Press RANGE and enter 10 for Xmin, 30 for Xmax, -20 for Ymin and 20 for Ymax. Using the TRACE command with the resulting graph (Figure 4.10c) will suggest locations of additional roots.

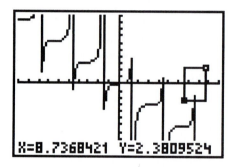

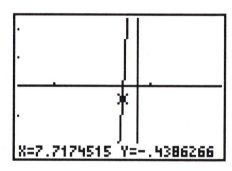

FIGURE 4.10a   FIGURE 4.10b

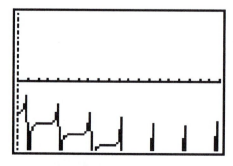

FIGURE 4.10c

Again, you can use ZOOM BOX to zoom in to get acceptable initial guesses for Newton's method. Continuing in this fashion, alternately translating the display to the right and then zooming in on the x-axis to see more detail, we can locate as many roots of f as needed. The first 5 (computed with our Newton's method program) are in the following table.

x_n	$f(x_n)$
4.493409458	0.000000000012
7.725251837	−0.000000000012
10.90412166	0.00000000004
14.06619391	−0.00000000048
17.22075527	−0.00000000057

At this point, you should verify the results in the preceding table and try to find the next largest positive root. Notice that one can see from Figure 4.8 about where the roots should be. However, we need to use the TI-81/85 plots to obtain acceptable initial guesses for Newton's method.   ■

You should recognize the interplay between the theory, the computation and the graphics. The theory of equations, by itself, is insufficient for finding roots or even for determining how many roots there may be. On the other hand, as we have seen, we cannot go around blindly stuffing guesses into Newton's method in the hope of finding a root, let alone all of the roots. We should emphasize that, in solving practical problems, we need to take care to use all of the information at our disposal: theory, computation, freehand graphs and numerically generated graphs, such as those produced by the TI-81/85.

MULTIPLE ROOTS

You may recall seeing the following definition.

Definition A function f is said to have a *root of multiplicity* n at $x = a$ if we can write

$$f(x) = (x - a)^n g(x)$$

where $\lim_{x \to a} g(x)$ exists and is nonzero.

Example 3. A Function with a Multiple Root

The function $f(x) = (x-2)^3(x^2+1)$ has a root of multiplicity 3 at $x = 2$.   ∎

All of the rootfinding methods which we have discussed encounter difficulties when used to look for a root which has a multiplicity greater than 1.

Example 4. An Example where Newton's Method Is Very Slow

Locate a root of $f(x) = (x-3)^2$. First note that f has only one root, a root of multiplicity 2 at $x = 3$. Also, note that Bisections cannot be used to locate this root, since $f(x) \geq 0$ for all x [i.e., $f(x)$ is never negative]. We apply Newton's method, using the initial guess $x_0 = 2$. The method works, but is unusually slow (more than the usual 5 or 6 iterations are required):

n	x_n	$x_n - x_{n-1}$
1	2.5	.5
2	2.75	.25
3	2.875	.125
4	2.9375	.0625
5	2.96875	.03125
⋮	⋮	⋮
⋮	⋮	⋮
26	2.999999985	.000000015
27	2.999999993	.000000008
28	2.999999996	.000000003

From the preceding table, we can see that, for the problem at hand, Newton's method took many more steps than it usually does. Looking down the last column in the table, we can see that each successive value is about half the preceding value. This means that at each step of Newton's method, we move only about half as far (ideally toward a root) as in the preceding step. This is very slow for Newton's method and is more like the convergence of the method of Bisections. The Secant method performs equally poorly for this problem. (Try it!)   ∎

REMARK: You might want to blame the poor performance of Newton's method on having made a poor choice of the initial guess. However, a better initial guess will not improve the situation. This behavior is typical of convergence to a root whose multiplicity is greater than 1. The interested reader is referred to more advanced texts on the subject of numerical analysis for a more complete exposition (e.g., Burden and Faires, *Numerical Analysis* , 4th edition).

Unfortunately, slow convergence is not the only problem which we face when there are roots of multiplicity greater than 1. Consider Example 5.

Example 5. An Example where Newton's Method Has Poor Accuracy

Find the roots of $f(x) = x^4 - 5x^3 + 6x^2 + 4x - 8$. From the initial TI-81 graph (see Figure 4.11), it is obvious that there is a positive root, but its location is not clear. A good guess might be that it lies around $x = 2$ or $x = 2.21$. So, we take as our initial guess the average of these two values, $x_0 = 2.105$. Using Newton's method, we obtain the following table from the TI-81.

n	x_n	$f(x_n)$
1	2.070390127	0.0010708504
2	2.047104695	0.00031847837
3	2.031483625	0.000094603984
⋮	⋮	⋮
⋮	⋮	⋮
16	2.000178802	0.00000000001
17	2.000144051	0.000000000006
18	2.000111924	0.0

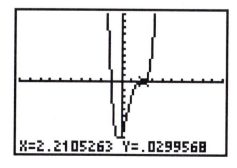

FIGURE 4.11

The TI-85 gets similar but slightly different results. Again starting with $x_0 = 2.105$, Newton's method terminates when $x_{22} = 2.00003853268$ and $f(x_{22}) = 0$.

This would at first seem to be a rather unremarkable example (except possibly for the noticeable discrepancy between the TI-81 and the TI-85). Certainly, Newton's method took many more steps than usual, but we've already seen that slow convergence can be caused by a root of multiplicity greater than 1. Quite naturally, then, we would conclude that $x = 2.000111924$ is an approximate root (maybe even a root of multiplicity greater than 1). After all, Newton's method converged to this value and further steps yielded no further progress. If this isn't enough evidence, the value of f at $x = 2.000111924$ is reported to be 0.0. How much more evidence do we want, anyway?

At this point, it might be useful to notice that the given polynomial factors:

$$f(x) = x^4 - 5x^3 + 6x^2 + 4x - 8 = (x+1)(x-2)^3$$

Thus, the only roots are $x = -1$ and $x = 2$, with the latter being of multiplicity 3. Our suspected approximate root of $x = 2.000111924$ is then seen to be accurate only to the first three decimal places. This is poor performance at best.

So, what went wrong with this application of Newton's method? Without getting into too many details, we can explain this as follows. If x is "close" to 2, $(x-2)$ will be close to zero, but the factor $(x-2)^3$ will be *very* close to zero. In the present case, for $x = 2.000111924$, $(x-2) = .000111924$ and $(x-2)^3 = 1.4E - 12$. ∎

What, then, is the moral of this story? Should we learn that in the case of roots of multiplicity greater than 1, we should demand a somewhat smaller tolerance? Certainly, we cannot expect to get better than $f(x_n) = 0$ and $x_n - x_{n+1} = 0$. Perhaps we should be wary of all rootfinding problems involving multiple roots. The flaw in these suggestions is that we would need to know (in advance) when the root that we are seeking has a multiplicity greater than 1. In practice, this can be done only by observing the slow convergence of our rootfinding scheme.

We should beware of problems with inordinately slow convergence. This generally suggests a multiple root, and that spells trouble. We need to realize, too, that because of the accuracy limitations of the TI-81/85 (usually about 10 digits for the TI-81 and 12 for the TI-85) we will not, in many cases, be able to obtain even moderate accuracy in the root. It is interesting to note that program BISECT also performs poorly on this problem. The culprit here is an inability to compute function values accurately for x very close to 2 [e.g., the TI-81 calculates $f(2.000048828)$ to be negative, while the TI-85 calculates $f(2.00005371094)$ to be zero].

At this point, it might be interesting to see how well the built-in TI-85 Solver does on this problem. With the initial guess of 2.105 [enter the expression x∧4− 5x∧3 + 6x∧2 + 4x − 8, set exp to 0, set x to 2.105 and, with the cursor still on the x-value line, press SOLVE (F5)] the Solver gives the approximate root $x =$ 1.9999660178825, with the advisory message "lft-rt=0" indicating that the machine thinks that it has found a root exactly. Of course, this answer is only marginally better than the unacceptable answer found by our Newton's method program.

Notice that there is an even greater concern with using the Solver to solve such problems. Since we cannot observe the calculation in progress (as we can with our Newton's method program) we have no idea when something may be wrong. On the contrary, the advisory message "lft-rt=0" (i.e., the function value at the reported root is 0) serves to convince us that everything is just fine and that we have found an accurate approximation to a root. For this reason, we caution against using the Solver alone to find roots. When there are multiple roots, the Solver (and hence also the user) can be easily fooled into making an incorrect conclusion. We suggest that you use the Newton's method and Secant method programs given in this chapter and pay close attention to the progress of the calculation and not just to the final answer.

There are some things which can be done to improve the situation. However, a complete treatment of these methods is beyond the scope of the present work. We will give some hints in the exercises, but for a thorough treatment, the interested reader is referred to a text on numerical analysis. The real lesson for us here is to learn *caution* in solving for roots numerically. In practice, you must use a great deal of care, especially when a root of multiplicity greater than 1 is detected.

Exercises 4.2

In exercises 1-6, rewrite the equation in the form $f(x) = g(x)$ and use graphics to determine how many roots there are.

1. $\cos x^2 + x = 0$

2. $\sin x^2 + x^3 - 2x^2 + 1 = 0$

3. $x^8 + 3x^6 + 4x^2 - 4 = 0$

4. $(x^2 - 1)^{2/3} + 3x - 1 = 0$

5. $e^{-x} + x^2 - 1 = 0$

6. $3e^{-x}\cos(x - 1) - x^2 + 2x - 2 = 0$

In exercises 7-8, there are an infinite number of roots. Use graphics and a rootfinder to determine the three smallest positive roots.

7. $\sec x - x = 0$ 　　　　　　　　　　　　　　8. $e^{-x} = \tan x$

In exercises 9-12, use graphics and a rootfinder to determine *all* the roots. Based on the rate of convergence, which roots do you suspect are multiple roots?

9. $x^4 - 12x^2 + 32 = 0$ 　　　　　　　　　　10. $x^5 - x^4 - 10x^3 + 10x^2 + 25x - 25 = 0$

11. $x^5 - 10x^4 - 2x^3 + 20x^2 - 3x + 30 = 0$ 　　12. $x^4 + 2x^3 - 6x^2 - 14x - 7 = 0$

In exercises 13-14, $x = 1$ and $x = 2$ are roots. In exercises 15-16, $x = 0$ is a root. Based on the convergence of Newton's method, determine which are multiple roots.

13. $x^4 - 2x^3 - 3x^2 + 8x - 4 = 0$ 　　　　14. $x^4 - 7x^3 + 18x^2 - 20x + 8 = 0$

15. $x \sin x = 0$ 　　　　　　　　　　　　　16. $x(\cos x - 1) = 0$

17. In this exercise, we look at an alternative stopping rule for the Newton's method algorithm. Let $f(x) = x^4 - x^3 - 3x^2 + 5x - 2 = (x - 1)^3(x + 2)$.
    (a) Execute Newton's method with $x_0 = 1.5$. Stop when $|x_{n+1} - x_n| < .0001$.
    (b) Repeat part (a) but stop when $|f(x_n)| < .0001$.
    (c) Compare the number of steps executed and the accuracy.

18. All of our examples so far have had relatively small roots. Special problems may occur if a root is large. Consider $f(x) = (x - 400)^2(x + 1) = x^3 - 799x^2 + 159200x + 160000$.
    (a) Execute Newton's method with $x_0 = 300$. Stop when $|x_{n+1} - x_n| < .0001$.
    (b) Repeat part (a) but stop when $|x_{n+1} - x_n| < .0001|x_{n+1}|$.
    (c) Compare the number of steps executed and the accuracy. Under what circumstances might criterion (b) be more appropriate than criterion (a)?

In exercise 19, we will see one way to speed up the convergence of Newton's method in the case of a multiple root. Use this method to solve exercises 20-22.

19. Show that if f is a polynomial with a root c of multiplicity n then c is a root of multiplicity 1 of f/f'. In this case, Newton's method would converge rapidly to c if f were replaced by f/f'. HINT: $f(x) = (x - c)^n g(x)$.

20. $x^4 - 5x^3 + 6x^2 + 4x - 8$ with $x_0 = 2.15$ (see Example 5).

21. $x^4 - x^3 - 3x^2 + 3x - 6$, $x_0 = 1.5$ 　　　22. $x^3 - x^2 - x + 1$, $x_0 = 1.2$

EXPLORATORY EXERCISE

Introduction

We have seen that Newton's method exhibits slow convergence to roots of multiplicity 2 or more. In exercises 19-22, we saw a messy way of speeding up the rate of convergence. Yet, we have never precisely said what we mean by *rate of convergence*. The stopping criterion we have used in our rootfinding methods is the difference $|x_{n+1} - x_n|$ between successive approximations. We use the quantity $\Delta_k = x_k - x_{k-1}$ to define the rate of convergence.

Problems

Start by running Newton's method with $x_0 = 1.5$ on the following examples while computing $\dfrac{\Delta_{k+1}}{\Delta_k}$ after steps 2, 3, ...

(a) $(x - 1)(x + 2)^3 = x^4 + 5x^3 + 6x^2 - 4x - 8$

(b) $(x - 1)^2(x + 2)^2 = x^4 + 2x^3 - 3x^2 - 4x + 4$

(c) $(x - 1)^3(x + 2) = x^4 - x^3 - 3x^2 + 5x - 2$

(d) $(x - 1)^4 = x^4 - 4x^3 + 6x^2 - 4x + 1$

Conjecture a value for $r = \lim\limits_{k \to \infty} \dfrac{\Delta_{k+1}}{\Delta_k}$ in cases (a)-(d). If r exists and is nonzero the method is said to *converge linearly*. Based on your calculations, formulate a hypothesis relating r to the multiplicity of the root. According to your hypothesis, what happens to the rate of convergence as the multiplicity of the root increases? Use Newton's method and your hypothesis to guess the multiplicity of the root $x = 0$ in the following cases.

(e) $x \sin x$ (f) $x \sin x^2$ (g) $x(x^x - 1)$

Further Study

The rate of convergence is a standard topic in numerical analysis. Introductory texts such as *Numerical Analysis*, 4th edition, by Burden and Faires have nice treatments of this and other numerical analysis topics. Other numerical analysis texts of interest are referenced in the Bibliography beginning on page 260.

4.3 Extrema and Applications

Everywhere we turn in business and industry today, we find someone asking questions like: "What's the least amount of time sufficient for completing this job?" or "What's the most money we can make on this investment?" or "What's the least amount of material that must be used to fabricate this device?" These questions are examples of what are called *optimization problems*, specifically *maximum/minimum* problems. In this section, we discuss some practical aspects of solving these problems. We start by pointing out that most problems encountered in a typical calculus textbook, of necessity, have solutions which are roots of a quadratic or, at worst, a cubic polynomial. Further, the solutions most often turn out to be integers.

As you might guess, it is rare that in a real world problem, we would be so fortunate as to be presented with a quadratic polynomial with integer roots. Yet, it is not surprising that most textbook problems are so limited. To be sure, our facility for finding roots by pencil-and-paper methods is confined (with few exceptions) to low degree polynomials.

We should recognize, too, that there are few among us who do not instantly frown when the solution of a (textbook) problem starts to involve numbers other than integers. Most of us, unfortunately, have been trained to do just that. At the same time, it is precisely this type of problem (messy, user-hostile ones) which you will be facing when you apply calculus to almost any real world problem. As users of calculus, we must come to the point where we *expect* to get messy-looking answers and are surprised at the odd instance when we get an integer answer. The power of the TI-81/85 is ideally suited for dealing with these problems, using the rootfinding skills developed in the last two sections.

REVIEW OF ABSOLUTE EXTREMA

Recall the following standard definition from elementary calculus.

Definition For x_0 in $[a, b]$, we call $f(x_0)$ the *absolute maximum* of f on the interval $[a, b]$ if $f(x_0) \geq f(x)$ for all x in $[a, b]$. $f(x_0)$ is the *absolute minimum* of f on $[a, b]$ if $f(x_0) \leq f(x)$ for all x in $[a, b]$. In either case, we call $f(x_0)$ an *absolute extremum*.

It should be fairly evident why someone would be interested in finding extrema. Simple examples are everywhere: Business managers want to maximize profits, while minimizing costs. Engineers are interested in maximizing the amount of energy which can be obtained from a barrel of oil and in minimizing the amount of raw material required to manufacture a given product. Similar examples abound in every branch of science, engineering, business and economics.

The first question that we might ask regarding the mathematics is whether every function has absolute extrema. The answer is no, but we can say the following.

Theorem 4.3 (Extreme Value Theorem) Suppose that f is a continuous function on the closed interval $[a, b]$. Then, f has both an absolute maximum and an absolute minimum on $[a, b]$.

Certainly, it is very comforting to know that for a large group of functions (the set of continuous functions), there will always be absolute extrema on a closed interval. But, how do we find what those values are? In Chapter 1, we drew a graph of the function and tried to read from the graph what appeared to be the extrema. Naturally, this process is just a bit too crude, even if the graph is computer generated (such as by those produced by the TI-81/85). There is a much more precise way to examine these problems. We can use the rootfinding methods developed in sections 4.1 and 4.2 to find approximations of extrema. First, we need the following definition.

Definition A number x_0 in the domain of a function f is called a *critical value* of f if $f'(x_0) = 0$ or if $f'(x_0)$ is undefined.

We can now state the main tool used for locating absolute extrema.

Theorem 4.4 Suppose that f is a continuous function on the closed interval $[a, b]$. Then, if $f(c)$ is an absolute extremum of f on $[a, b]$, c must be an endpoint (a or b) or a critical value.

NOTE: This result says that if we want to find the absolute extrema of a continuous function on a closed interval $[a, b]$, then we need only locate all of the critical values in (a, b) and simply *compare* the value of f at the two endpoints and at each of the critical values. The largest of these numbers will be the absolute maximum; the smallest will be the absolute minimum. (That at least *sounds* easy, doesn't it?)

Example 1. Finding Extrema of a Polynomial

Find the absolute extrema of $f(x) = x^3 - 3x^2 - 9x + 7$ on the interval $[-2, 5]$. Here, $f'(x) = 3x^2 - 6x - 9 = 3(x - 3)(x + 1)$. Thus, the critical values are the roots of $f'(x)$: $x = 3$ and $x = -1$. Since $f'(x)$ is a polynomial, it is defined everywhere and, hence, the only critical values of f are roots of f'. (The authors, of course, realize that this is another cooked-up textbook problem with integer roots, but it will serve as a good illustration of the procedure before we turn to the more messy and realistic problems to follow.)

Now, we compare the value of f at the endpoints and at the critical values to determine the relative extrema:

$$f(-2) = 5, \quad f(5) = 12, \quad f(-1) = 12 \quad \text{and} \quad f(3) = -20$$

Obviously, the absolute maximum of f on $[-2, 5]$ is 12 (this occurs at both $x = -1$ and $x = 5$) and the absolute minimum is -20 (this occurs at $x = 3$). ∎

Example 2. Extrema of a Function with a Fractional Exponent

Find the absolute extrema of $f(x) = (x^2 - 4)^{2/3}$, on the interval $[-1, 3]$. Here, $f'(x) = \dfrac{2}{3}(x^2 - 4)^{-1/3}(2x) = \dfrac{4x}{3(x^2 - 4)^{1/3}}$. In this case $f'(x) = 0$ only for $x = 0$. Further, $f'(x)$ is undefined whenever the denominator is zero, i.e., for $x = -2$ and $x = 2$ (both of which are in the domain of f). However, $x = -2$ is not in the interval under consideration. So, we need only compare:

$$f(-1) = (-3)^{2/3} = 2.080083823$$
$$f(3) = (5)^{2/3} = 2.924017738$$
$$f(0) = (-4)^{2/3} = 2.5198421$$
$$f(2) = (0)^{2/3} = 0$$

Clearly, the absolute maximum is $f(3) = 2.924017738$, while the absolute minimum is $f(2) = 0$. Check these results yourself. In doing so, you will notice that the TI-81/85 (like most calculators) has a problem when asked to compute certain fractional powers of negative numbers. In these cases, the user must compute the indicated power of the absolute value of the given number and then manually adjust the sign, if necessary. Alternatively, this particular function may be rewritten as $f(x) = [(x^2 - 4)^2]^{1/3}$. ∎

Now that we have reviewed the procedure for locating absolute extrema, the only remaining questions are computational ones (i.e., how do we actually find the critical values?). We will deal with these questions next.

How often in practice does one run into an extrema problem where the critical values are integer roots of a quadratic polynomial? If we were to answer "occasionally" we might still be guilty of exaggeration. Real world problems are rarely very pleasant and almost always require some computing to solve. The TI-81/85 is very well suited for solving many such problems. We exhibit here some examples that are typical of the type of problems encountered in applications. Pay particular attention to the interplay between the graphing, the analysis and the computation. As we'll see, no one of these three tools is sufficient for solving extrema problems in practice, but together they form a powerful combination.

Example 3. An Extrema Problem with Ugly Numbers

Find the absolute extrema of $f(x) = x^4 + 3x^3 - 5x^2 - 2x + 10$ on the interval $[-4, 2]$. It is always best to first get a rough idea of where the extrema might be from a graph of the function f. In the initial graph produced by the TI-81, we see that much of the graph does not fit on the display (see Figure 4.12a). Essentially, we need to compress the height of the graph, in order to fit it in the display window. At the same time, we are interested only in that portion of the graph on the interval $[-4, 2]$, while the standard display will show much more.

First, press $\boxed{\text{RANGE}}$ and set Xmin to -4 and Xmax to 2. Next, to compress the height, press $\boxed{\text{ZOOM}}$ 4 (Set Factors) on the TI-81 or $\boxed{\text{ZOOM}}$ $\boxed{\text{MORE}}$ $\boxed{\text{MORE}}$ $\boxed{\text{ZFACT}}$ on the TI-85. Set XFact to be 1 and YFact to be 4. Then press $\boxed{\text{ZOOM}}$ 3 (Zoom Out) on the TI-81 or $\boxed{\text{ZOOM}}$ $\boxed{\text{ZOUT}}$ on the TI-85 and press $\boxed{\text{ENTER}}$. This will leave the x-range fixed, but show more of the y-axis (see Figure 4.12b). You can now clearly see the entire graph over the interval of interest.

Although we should not completely trust the picture, it suggests that the absolute maximum is at the endpoint $x = 2$, and the absolute minimum is at a relative minimum, near $x = -3$. [Recall that f has a *relative minimum* at $x = c$ if and only if $f(c) \leq f(x)$, for all x in some open interval (a, b) containing c.] To verify these conclusions and to make the values more precise, we must find the critical values and compare the values of the function at the endpoints and the critical values.

Notice that $f'(x) = 4x^3 + 9x^2 - 10x - 2$. This has no obvious factorization (at

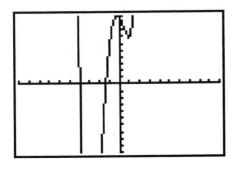

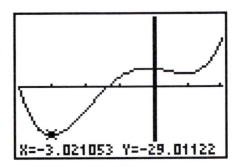

FIGURE 4.12a　　　　　　　　　　　FIGURE 4.12b

least not one obvious to the authors) and hence, we must rely on numerical methods to find the roots. First, use the graphics routines to graph $f'(x)$ to get an idea of where any roots might be. You need not reset the graphing parameters first, since the x-range is already set for the interval under consideration (see Figure 4.13a). You might also want to superimpose the graphs of f and f', in order to observe the connection between extrema and zeros of the derivative (see Figure 4.13b). From either graph, one can clearly see that there are three roots: one near $x = -3$, one near $x = -.2$ and one near $x = 1$. Since we know that a cubic polynomial can have at most three roots, there is no need to look any further.

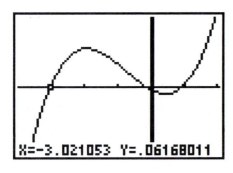

FIGURE 4.13a　　　　　　　　　　　FIGURE 4.13b

Using our Newton's method program [looking for roots of $f'(x)$] with these initial guesses, we obtain the following results:

x_0	Approximate Root x	$f'(x)$
-3.0	-3.022417859	-0.00000000004
-0.16	-0.1746723457	-0.000000000001
1.0	0.9470902049	-0.000000000005

We now need only compare the values of f at the endpoints and at the above three critical values. We have:

$$f(-4) = 2$$
$$f(2) = 26$$
$$f(-3.022417859) = -29.01125988$$
$$f(-.1746723457) = 10.18173545$$
$$f(.9470902049) = 6.974055679$$

Obviously, the absolute maximum occurs at $x = 2$, and the absolute minimum occurs at $x = -3.02241785918$, both as expected from the graph of f.   ■

We should note that in Example 3 the critical values are the roots of a cubic equation which you probably cannot see how to factor. There is, of course, a formula for finding the roots of a cubic equation, although it is rather cumbersome to use. We chose instead to approximate the roots numerically. This is more instructive, of course, since in general there are no formulas for finding roots exactly. You should also note that, in this example, we knew in advance that there could be no more than 3 critical values. In general, we will have no idea of how many critical values to expect. We will therefore need to carefully search for critical values. Example 4 is typical of the general situation.

Example 4. Extrema of a Non-Polynomial Function

Consider $f(x) = \cos x^2 + x^3 - 2x$ on the interval $[-1, 2]$. As in Example 3, after drawing the graph in the default window [ZOOM 6 (Standard) on the TI-81 or ZOOM ZSTD on the TI-85] match the x-range to the interval of interest (press RANGE and set Xmin to -1 and Xmax to 2). Pressing GRAPH produces the graph in Figure 4.14a (taken from the TI-81). While all of the graph fits on the screen, the graph is too compressed in the y-direction to see the precise behavior. The solution is to zoom in some. Press ZOOM 4 (Set Factors) on the TI-81 or ZOOM MORE MORE ZFACT on the TI-85. Then, set XFact to be 1 and YFact to be 4. (If these have not been changed since the last example, you may skip this step.) Finally, press ZOOM 2 (Zoom In) (or ZOOM ZIN on the TI-85) and ENTER . The behavior on the interval of interest should now be clear (see Figure 4.14b).

The graph seems to indicate that the absolute maximum occurs at $x = 2$, while the absolute minimum occurs around $x = 1.2$. Also, there seems to be a relative

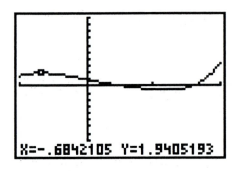

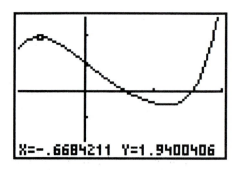

FIGURE 4.14a   FIGURE 4.14b

maximum around $x = -.7$. [Recall that f has a *relative maximum* at $x = c$ if and only if $f(c) \geq f(x)$, for all x in some open interval (a, b) containing c.] From this, we expect to find 2 critical values in the interval $(-1, 2)$.

Next, to make sure that we find all of the critical values, draw a graph of $f'(x) = -2x \sin(x^2) + 3x^2 - 2$. Again, there is no need to first adjust the graphing window, since it is already set to the proper interval of x-values. From Figure 4.15, we see that there are roots near $x = -.7$ and near $x = 1.2$.

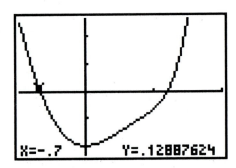

FIGURE 4.15

Using our Newton's method program with these starting values, we get:

x_0	Approximate Root x	$f'(x)$
-0.7	-0.6809207671	0.0
1.2	1.212981165	$-2E{-}12$

So, we have found approximations to 2 critical values, where we had expected only 2 critical values. Further, these 2 values are located near where we had expected to find the extrema. Thus, we have no reason to search for any other critical values, in this case. It remains only to compare the values of the function f at the endpoints and at the critical values:

$$f(-1) = 1.540302306$$

$$f(2) = 3.346356379$$

$$f(-.6809207671) = 1.940555255$$

$$f(1.212981165) = -.5419658158$$

We can now read off the extrema. The absolute maximum occurs at $x = 2$ and the absolute minimum occurs at $x = 1.212981165$, both as expected. You should not underestimate the importance of checking that the computed extrema correspond to those expected from the graph of $f(x)$. If they do not, then you should return to the graph and manipulate it so as to determine what you missed the first time (e.g., a missed critical value or an incorrect location for the relative extrema).   ∎

By now, you should realize that we should be able to find the approximate locations for the absolute extrema of a continuous function, just by drawing a sufficiently detailed graph. However, to find the precise locations and to find the precise maximum and minimum values, we must rely on the results of our numerical rootfinding schemes and a comparison of function values. This process exhibits the interplay between the mathematical analysis, numerical computation and graphics typical of so many practical problems and which we have already seen in several other contexts. We provide one final example of such a problem, one where the graphics are not so easy to work with.

Example 5. Extrema of a Difficult Polynomial

Find the absolute extrema of $f(x) = x^6 - 7x^4 + 3x^3 - 5x + 1$ on the interval $[-3, 3]$. First draw the graph in the default window [press $\boxed{\text{ZOOM}}$ 6 (Standard) on the TI-81 or $\boxed{\text{ZOOM}}$ $\boxed{\text{ZSTD}}$ on the TI-85]. Then, set the x-range to match the interval of interest (press $\boxed{\text{RANGE}}$ and set Xmin to -3 and Xmax to 3; see Figure 4.16a). Since much of the graph does not fit in the current window, we zoom out in the y-direction, by pressing $\boxed{\text{ZOOM}}$ 3 (Zoom out) and $\boxed{\text{ENTER}}$ on the TI-81 or by pressing $\boxed{\text{ZOOM}}$ $\boxed{\text{ZOUT}}$ $\boxed{\text{ENTER}}$ on the TI-85. (First, if you have adjusted the Zoom Factors, make sure that you reset them to XFact = 1 and YFact = 4.) This will produce the graph in Figure 4.16b.

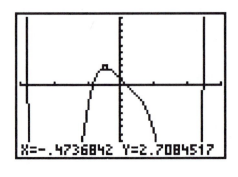
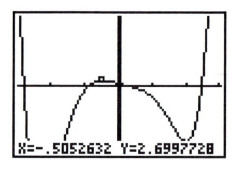

FIGURE 4.16a   FIGURE 4.16b

You should notice the relative maximum near $x = -.5$, but the behavior on the interval $[-2.7, -1.6]$ is still not clear. To remedy this, we zoom out once more. We can finally see where the graph bottoms out (see Figure 4.17). It appears that there is an absolute minimum near the point $(-2.3, -72)$ and that the absolute maximum will occur at the endpoint, $x = 3$. Now that we have an idea of what we are looking for, we turn to finding the critical values.

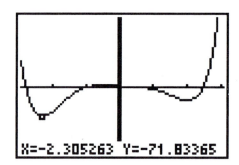

FIGURE 4.17

Graphing the derivative, $f'(x) = 6x^5 - 28x^3 + 9x^2 - 5$, we look for zeros in the interval of interest, $(-3, 3)$. (Once again, it may help to draw superimposed graphs of f and f'.) Pay particular attention to locating any critical values near the suspected relative extrema seen in the graph of f. If we draw the graph of f' without resetting the graphics window, we get the graph in Figure 4.18a. From this, you can clearly see that there are roots of f' near $x = -2.3$ and near $x = 2$. There also seems to be a root somewhere on the interval $[-1, 1]$. To get a better idea of where this root might be, we zoom in on a small box enclosing that part of the graph (see Figure 4.18b). You should now be able to clearly see that there is a root of f' near $x = -.48$. Further, you should be convinced from the graph in Figure 4.18a that we have indeed found all of the roots of f' on the interval $[-3, 3]$.

X=-2.305263 Y=-4.771154   X=-.4696953 Y=-.2502449

FIGURE 4.18a   FIGURE 4.18b

So, from some routine work with the graphics, we have three guesses for critical values, -2.3, $-.48$ and 2. Using our Newton's method program [for roots of $f'(x)$], with these initial guesses, we get the following results.

x_0	Approximate root x	$f'(x)$
-2.3	-2.291715813	-0.00000000007
-0.48	-0.4793442703	-0.000000000002
2.0	2.005501386	0.0

Finally, we compare the values of f at the critical values and at the endpoints:

$$f(-3) = 97$$
$$f(3) = 229$$
$$f(-2.291715813) = -71.86575095$$
$$f(-.4793442703) = 2.708871138$$
$$f(2.005501386) = -33.00275965$$

We can now see that the absolute maximum occurs at $x = 3$, and the absolute minimum at $x = -2.291715813$, as expected from the graph of f.   ∎

Of course, the reason that we are interested in solving extrema problems is that they occur quite naturally in applications. We offer Example 6 as an illustration of the typical applied max/min problem, where the solution cannot be easily found through means of elementary algebra.

Example 6. Applied Maximum/Minimum

A city would like to build a new section of superhighway to link an existing bridge with another highway interchange, lying 8 miles to the east and 8 miles to the south of the bridge. Unfortunately, there is a 5-mile-wide stretch of marsh land which must be crossed (see Figure 4.19). Given that the highway costs 10 million dollars per mile to build over marsh and 7 million dollars per mile to build on dry land, how far to the east of the bridge should the highway be at the point where it crosses out of the marsh?

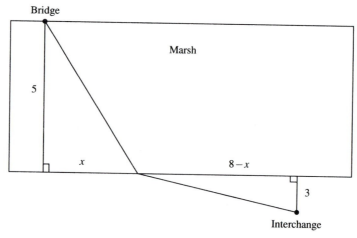

FIGURE 4.19

As with any applied max/min problem, you should first draw a picture and label the appropriate variables, as we have done in Figure 4.19. Let x represent the distance in question (marked in Figure 4.19). Then the total cost of the project (in millions of dollars) is

$$\text{Cost} = 10 * \text{distance across marsh} + 7 * \text{distance across land}$$

In Figure 4.19, there are obviously two right triangles. Using the Pythagorean theorem, we find that the cost is given by

$$C(x) = 10(x^2 + 25)^{1/2} + 7[(8 - x)^2 + 9]^{1/2}$$

Store this function in the "Y=" list as Y1. From the picture, it is easy to see that $0 \le x \le 8$.

So, we would like to find the minimum value of C on the interval $[0,8]$. First, to get an idea of where the minimum might be, draw a graph of $y = C(x)$. If we (as usual) reset the graphics window and then set the x-range to match the interval under consideration, the values of the function are obviously all off the scale (no points at all are plotted). In applications, the first step is often to think about what reasonable values of the variables are. We can use the function evaluation program FEVAL (given in Chapter 1) to help us with the y-range. We find that $C(0) \approx 110$, $C(4) \approx 99$ and $C(8) \approx 115$. With this as a guide, we set Ymin=90 and Ymax=120. The TI-85 graph is given in Figure 4.20. From this graph, we can see that the minimum value seems to be located between $x = 3.3$ and $x = 3.8$.

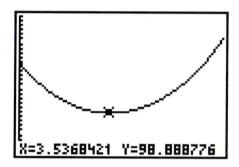

X=3.5368421   Y=98.888776

FIGURE 4.20

To find the minimum value precisely, we will first need to find any critical values. The derivative of C is

$$C'(x) = 10x(x^2 + 25)^{-1/2} - 7(8 - x)[(8 - x)^2 + 9]^{-1/2}$$

Notice that the only critical values occur where $C'(x) = 0$. (Why is that?) We now look for the roots of $C'(x)$. To get an idea of where these might be, we first draw a graph of $C'(x)$. Since we are now looking for roots and the last plotted graph was of a segment of $y = C(x)$ well up above the x-axis, we must first reset to the default graphics window. From this graph (Figure 4.21), we can clearly see that there is a root near $x = 3.6$ and that there are no other roots on the interval of interest.

Note that we could use $x = 3.6$ as an initial guess for Newton's method, but this would require that we compute the derivative of $C'(x)$. While this is not a monumental task (you could always use the differentiation routines of the TI-81/85

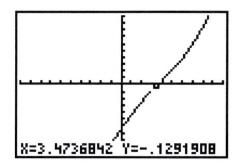

X=3.4736842 Y=-.1291908

FIGURE 4.21

to do this numerically for each needed value of x), it is simpler to use the Secant method in this case. Using the Secant method program, with the initial guesses $x_0 = 3$ and $x_1 = 4$, we get the approximate root

$$x_4 = 3.56005152 \qquad \text{where} \qquad C'(x_4) = .000000000001$$

Finally, compare the value of $C(x)$ at the endpoints and at the critical value:

$$C(0) = \$109,808,026.20$$
$$C(8) = \$115,339,811.30$$
$$C(3.56005152) = \$98,888,374.49$$

Thus, if the roadway is built so that $x = 3.56...$, this will result in a savings of more than 10 million dollars over cutting straight across the marsh and a savings of more than 16 million dollars over cutting diagonally across the marsh. ∎

The examples which we have given in this section together with the exercises to follow should give you the necessary tools for solving a wide variety of maximum/minimum problems found in applications. In solving such problems numerically, we urge caution, as always. You should be careful to check that the answer computed numerically corresponds to the solution expected from the graph of the function being maximized or minimized. If it does not, then further analysis is needed. Perhaps a refined graph of the function will shed some light on the problem or perhaps a critical value was missed in the graph of $f'(x)$. You should also check that the solution makes physical sense, if possible. All of these multiple checks reduce the likelihood of error. A good problem-solver must be on guard all the time, for there are many traps to fall into.

Exercises 4.3

In exercises 1-14, do the following: (a) use graphics to predict the maximum and minimum of the function on the interval; (b) use graphics and a rootfinding method to approximate the critical points; (c) find the maximum and minimum of the function on the interval.

1. $x^3 - 6x^2 + 9x - 2$, $[-2,2]$

2. $x^3 - 6x^2 + 9x - 2$, $[0,4]$

3. $x^4 + 4x^3 - 6x^2 - 36x + 25$, $[-2,2]$

4. $x^4 + 4x^3 - 6x^2 - 36x + 25$, $[-4,0]$

5. $x^6 + 4x^4 - 3x^3 + 4x - 2$, $[-1,3]$

6. $x^6 + 4x^4 - 3x^3 + 4x - 2$, $[-3,0]$

7. $\sqrt{x^2 + 4} - \dfrac{x^2}{6} + 1$, $[-1,3]$

8. $\sqrt{x^2 + 4} - \dfrac{x^2}{6} + 1$, $[-4,1]$

9. $(x^2 - 1)^{2/3} - 2x + 1$, $[1,3]$

10. $(x^2 - 1)^{2/3} - 2x + 1$. $[-2,0]$

11. $x^2 \sin x - 2$, $[0,4]$

12. $x^2 \sin x - 2$, $[-4,0]$

13. $xe^{-x} + x^2$, $[0,2]$

14. $e^{-x} + x^2$, $[-4,4]$

15. Light travels at speed c in air and speed $.75c$ in water. Find x to minimize the time it takes light to get from point A in air to point B in water.

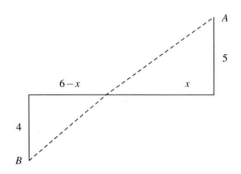

16. The points $A(0,1)$ and $B(0,-1)$ are within the circle $x^2 + y^2 = 4$. Consider the path starting at point A, reflecting off the circle and finishing at point B. Find the points (x,y) on the right half of the circle (that is, $x \geq 0$) which minimize and maximize the reflecting distance from A to B. It is interesting to note that light can follow both paths (usually, light only follows paths of minimum time).

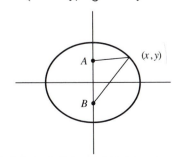

17. Washington needs to cross the Delaware to get to Trenton. Assume that the Delaware is 1 mile across and that Trenton is 2 miles inland and 10 miles downstream. If Washington's men can row at 3 mph and march at 4 mph, how far downstream should they row? How many minutes will they save compared to rowing straight across the river and then marching directly to Trenton?

18. The spinning of a clothes dryer causes it to vibrate as if being acted on by a downward force of f pounds, where $f = f_0 \sin wt$ for constants f_0 and w. Small springs and dampers may be used to reduce the vibrations. In studying the design of the machine (see **Raven**, *Mathematics of Engineering Systems*) an important quantity is $F = f_t/f_0$ where f_t is the amplitude of the force transmitted to the floor and f_0 is the amplitude of the vibrating force. This ratio has the form

$$F = \sqrt{\frac{1 + cb^2}{(1 - b^2)^2 + cb^2}}$$

where b is determined by the strength of the spring and c is determined by the amount of damping. For $c = .1$, find b to maximize F. For $c = .4$, find b to maximize F. (**HINT**: Find the maximum of F^2.) In practice, with $c = .1$ it is common to have $b = 4$. Explain why this differs from the value obtained above.

19. In sports where balls are thrown or hit, the ball often finishes at a different height than it starts at. Examples include a golf shot downhill and a basketball shot. In the diagram, a ball is released at an angle θ and finishes at an angle B above the horizontal (B can be negative for downhill trajectories). Neglecting air resistance and spin, the horizontal range is given by

$$R = \frac{2v_0^2 \cos^2 \theta}{g}(\tan \theta - \tan B)$$

where v_0 is the initial velocity of the ball and g is the gravitational constant. In the following cases, maximize R: (a) $B = 10$; (b) $B = 0$; (c) $B = -10$ (in degrees). Verify that $\theta = 45 + B/2$. [**HINT**: argue that you only need to maximize $\cos^2 \theta(\tan \theta - \tan B)$]. This result and other uses of calculus in sports can be found in *Sports Science* by Brancazio.

20. A ball is thrown from $s = b$ to $s = a$ ($a < b$) with initial speed v_0. Assuming that air resistance is proportional to speed, the time it takes the ball to reach $s = a$ is

$$T = -\frac{1}{c}\ln\left(1 - c\frac{b - a}{v_0}\right)$$

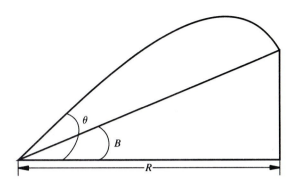

FIGURE FOR EXERCISE 19

where c is a constant of proportionality. A baseball player is 300 feet from home plate and throws a ball directly towards home plate with an initial speed of 125 ft/sec. Another player stands x feet from home plate and has the option of letting the ball go by or catching it and, after a delay of .1 seconds, throwing the ball towards home plate with an initial speed of 125 ft/sec. Take $c = .1$ and find x to minimize the total time for the ball to reach home plate. What, if anything, changes if the delay is .2 seconds?

21. For the situation in exercise 20, for what length delay is it equally fast to have a relay and not have a relay? Why do you suppose it is considered important in baseball to have a relay option?

22. Repeat exercises 20 and 21 if the second player throws the ball with initial speed 100 ft/sec.

23. For a delay of .1 second, find the value of the initial speed of the second player's throw for which it is equally fast to have a relay and not have a relay.

24. Repeat exercise 20 if the second player throws the ball with an initial speed of 120 ft/sec.

EXPLORATORY EXERCISE

Introduction

The theory developed in this section gives us a definite answer about max/min problems involving a continuous function of one variable on a closed interval. As you might expect, not all applications of interest fit into this category. In this exercise, we develop a technique for solving a different type of max/min problem. We will analyze an old problem known as the "farmer problem." A farmer standing at $(-2,0)$ needs to get water from a stream represented by $y = 6 - x$ and deliver the water to a cow at $(2,0)$. From which point on the stream should the farmer get the water to minimize the total walking distance?

Problems

We first find the solution graphically. The set of points for which the total walking distance is d is given by the ellipse

$$\frac{x^2}{(d/2)^2} + \frac{y^2}{(d/2)^2 - 4} = 1$$

This is called the *level curve* of the distance variable d. Trying the value $d = 7$, graph the top half of the ellipse and $y = 6 - x$ simultaneously and convince yourself that the farmer will have to walk more than 7 units. Repeat the above with $d = 11$ and convince yourself that the farmer can walk less than 11 units.

With $d = 9$ the line $y = 6 - x$ just barely passes inside the ellipse. This tells us that the farmer can walk less than 9 units (why?). More importantly, we are now close to the best point on the stream. By decreasing d slightly, we should be able to find an ellipse that touches the stream at only one point. This point would be the solution of our problem. Estimate this point.

We can find this point analytically, too. Note that at the optimal point, the ellipse and line are tangent to each other, and hence have the same slope. The slope of the line is -1, so we know 3 requirements for the optimal point: (a) it is on the line; (b) it is on the ellipse; (c) at this point the slope of the ellipse is -1. Translating these into equations, we get

$$y = 6 - x \qquad \frac{x^2}{a^2} + \frac{y^2}{a^2 - 4} = 1 \qquad \frac{x}{a^2} = \frac{y}{a^2 - 4}$$

where we have used $a = d/2$ for convenience. Find the optimal point.

Further Study

The geometry of the technique described above is the basis of a powerful result known as the Lagrange Multiplier Theorem. The theorem is normally stated in terms of a vector notation which simplifies calculations, but the principle is the same: at the optimal point, the level curve of the function to be optimized is tangent to the constraint curve (in the above example, the constraint curve is the stream $y = 6 - x$). This result is a fundamental part of the field of *calculus of variations*, which is typically a graduate course.

CHAPTER
5

Integration

5.1 Area and Riemann Sums

You are all familiar with the formulas for computing the area of rectangles, circles and triangles. We don't need to look very far to find good reasons for wanting to compute areas. For example, if you want to know how much grass seed you will need for your front yard, you'll need to find its area. The question of how to compute area is certainly much more profound than this example might make it seem, but this should serve to illustrate the point.

Most people's front yards are not perfect rectangles, circles or triangles. Does this mean that their yards don't have area? Certainly not, but the question remains as to how the area is to be computed. Notice that we've used the word *computed*. Areas are not measured directly, but rather are computed using some one-dimensional measurements and a formula or formulas.

What we need, then, is a more general description of area, one which can be used to find the area of almost any two-dimensional region imaginable. In this chapter, we will investigate the notion of the definite integral. At first, we will develop this as a tool for computing areas, but its usefulness goes far beyond this seemingly mundane question. It is, in fact, one of the central ideas of calculus. Our studies in this chapter will arm us with a powerful and flexible tool, one which has applications in a wide variety of fields.

RIEMANN SUMS

We start our exploration of area by looking at a simple example on the TI-81/85. First, graph the parabola $y = 2x - 2x^2$ using the graphing window described below. On the TI-81, set Xmin$=-4.8$, Xmax$=4.7$, Ymin$=-3.2$ and Ymax$=3.1$. On the TI-85, set xMin$=-6.3$, xMax$=6.3$, yMin$=-3.1$ and yMax$=3.1$. Recall that these values are adjusted through the Range menu. The TI-81 graph is shown in Figure 5.1a. We would like to find the area of the region bounded by the x-axis and the graph, as shaded in Figure 5.1b. (The shaded graph is produced using the Shade command, to be discussed shortly.) The region is clearly not a rectangle, a circle or a triangle, so we will look for an approximation of the area.

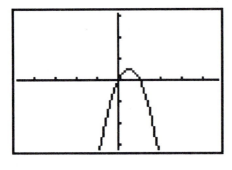

FIGURE 5.1a

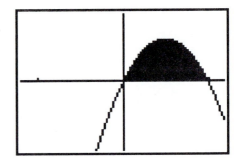

FIGURE 5.1b

Since $2x - 2x^2 = 0$ if $x = 0$ or $x = 1$, the region extends from $x = 0$ to $x = 1$.

Although we often think of the calculator as plotting points, it actually colors in small squares on your screen called *pixels*. If you look closely enough at your calculator display, you will see a picture similar to the colored-in graph paper shown in Figure 5.2.

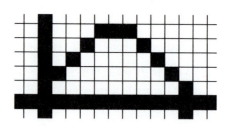

FIGURE 5.2

Move the cursor around the screen (using the free cursor, not the TRACE cursor) and notice that each step changes the x- or y-coordinate by .1 (this is our reason for choosing the particular viewing window given above). Therefore, each pixel represents a square of side .1. To estimate the area, then, we can add up the number of pixels in the region of interest and multiply the total by .01 (the area of each pixel). Notice that at $x = .1$, the graph is 2 pixels high, at $x = .2$ the graph is 3 pixels high and so on. Counting carefully, you should find

$$2 + 3 + 4 + 5 + 5 + 5 + 4 + 3 + 2 + 0 = 33$$

pixels, if you count the pixels representing the graph but not those representing the x-axis or the tick mark indicating $x = 1$. (Why don't we count both, or neither?) Our estimate of the area is then $(33)(.01) = .33$.

It can be shown that the exact area is $1/3$ (we'll see how later), so it would seem that we have found a fairly good estimate. However, we have not yet developed a general procedure for computing area. First, we will need a way of systematically obtaining better and better estimates. (You should notice that for larger areas you would quickly tire of counting pixels.)

We can improve on this pixel-counting strategy by tracing along the curve with the cursor. Specifically, if the cursor is located at the point $(.3, .4)$, then it is 4 pixels above the x-axis. Of course, the TI-81/85 allows us to follow the curve using the TRACE command. Note that at $x = .3$ the function value is given as $y = .42$. Locating the pixel at $(.3, .4)$ has the effect of rounding .42 down to .4. Perhaps if we use the actual function value of .42 we could get a better estimate of the area. For now, the important point is to see how the function values are related to the heights of the various columns of pixels. We exploit this in what follows.

Rather than thinking in terms of pixels, we can think of our area approximation in terms of rectangles sitting on the x-axis and fitted in under the graph. Notice that in the TI-81 graph (Figure 5.1a), at $x = .1$, the graph is 2 pixels high. The display then shows a rectangle of height .2 and width .1. Next to this, we see a rectangle of height .3 and width .1, and so on. Thus, Figure 5.3 is essentially the same as Figure 5.2, but with the rectangles shaded in.

Our estimate of the area, then, is the sum of the areas of the rectangles:

$$(.2 + .3 + .4 + .5 + .5 + .5 + .4 + .3 + .2 + 0)(.1)$$

where we have factored out the common width of .1. Using the TRACE command

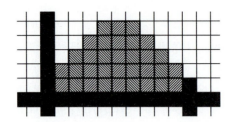

FIGURE 5.3

to get the actual function values at $x = .1, .2, \ldots$ we get the area estimate:

$$(.18 + .32 + .42 + .48 + .5 + .48 + .42 + .32 + .18 + 0)(.1)$$

which, coincidentally, is .33 again. Notice that we can rewrite this estimate as

$$(f_1 + f_2 + f_3 + f_4 + f_5 + f_6 + f_7 + f_8 + f_9 + f_{10})(.1)$$

where the notation used here suggests the relationship between the function values and the heights of the rectangles [i.e., $f_1 = f(.1)$, $f_2 = f(.2)$, $\cdots$, $f_{10} = f(1.0)$].

You might wonder what would happen if we took more and more rectangles of increasingly small width (i.e., smaller pixels). Since this means a higher resolution picture, we should obtain a better approximation of the area. Indeed, by generalizing the preceding process, we get a definition which is very useful for computing areas. We start by dividing the interval $[a, b]$ into n subintervals of equal length $\Delta x = (b - a)/n$. (This is called a *regular partition* of the interval.) For each subinterval $[x_{i-1}, x_i]$, $i = 1, 2, \ldots, n$, we choose any point c_i in $[x_{i-1}, x_i]$ (see Figure 5.4).

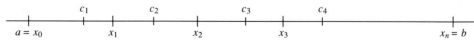

FIGURE 5.4

Definition The Riemann sum $R_n(f)$ of a function $f(x)$ corresponding to the above partition and the evaluation points $c_1, c_2, \ldots, c_n$ is

$$R_n(f) = [f(c_1) + f(c_2) + f(c_3) + \ldots + f(c_n)]\Delta x$$

Note that the value of a Riemann sum depends on the function, the choice of n, and the choice of the evaluation points.

Example 1. Computing Riemann Sums

Compute the Riemann sums with $n = 4$ and $n = 8$ for $f(x) = x^2$ on the interval $[1, 3]$, where for each $i = 1, 2, ..., n$, c_i is chosen to be the midpoint of the ith subinterval, $[x_{i-1}, x_i]$.

For $n = 4$, we find that $\Delta x = \dfrac{1}{2}$ and the subintervals that make up the partition are $\left[1, \dfrac{3}{2}\right]$, $\left[\dfrac{3}{2}, 2\right]$, $\left[2, \dfrac{5}{2}\right]$ and $\left[\dfrac{5}{2}, 3\right]$. The midpoints of the subintervals are then $c_1 = \dfrac{5}{4}$, $c_2 = \dfrac{7}{4}$, $c_3 = \dfrac{9}{4}$ and $c_4 = \dfrac{11}{4}$. Thus, we get

$$R_4(f) = [f(5/4) + f(7/4) + f(9/4) + f(11/4)]\frac{1}{2}$$

$$= [1.5625 + 3.0625 + 5.0625 + 7.5625]\frac{1}{2}$$

$$= 8.625$$

If the calculations seem tedious, you will be glad to know that we will shortly introduce a program to compute these sums automatically.

For $n = 8$, you should verify that the Riemann sum is

$$R_8(f) = [f(9/8) + f(11/8) + f(13/8) + ... + f(23/8)]\frac{1}{4} = 8.65625$$

∎

In Figures 5.5a and 5.5b, we see the rectangles corresponding to the Riemann sums R_4 and R_8, respectively. Based on these figures, we would expect R_8 to be the better approximation of the actual area under the curve. (Why is that?) In fact, you should convince yourself that the larger n is, the better the approximation R_n should be.

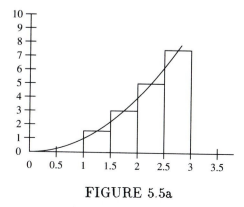

FIGURE 5.5a

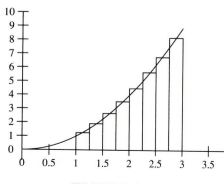

FIGURE 5.5b

The following program computes $R_N(f)$ for a given integer N, an interval $[A, B]$ and a function f stored as Y1. The parameter R, $0 \le R \le 1$, lets us vary the evaluation points from the left endpoint of each subinterval ($R = 0$) to the right endpoint ($R = 1$), to anything in between ($0 < R < 1$). In Example 1 above, we used midpoint evaluations, which corresponds to the choice $R = .5$. The first program listed will work on both the TI-81 and TI-85, but we recommend that TI-85 users type in the second program. We suggest that TI-81 users save this program as program R. (Program TRAP below uses program RIEM, referring to it by the name PrgmR.) Note that the boxed commands are available in TI-81 menus. In particular, you will press $\boxed{\text{PRGM}}$ 1 (Lbl), $\boxed{\text{PRGM}}$ 2 (Goto), $\boxed{\text{PRGM}}$ 3 (If), $\boxed{\text{TEST}}$ 1 (=) and $\boxed{\text{TEST}}$ 5 (<) while entering this program.

:RIEM   :0$\boxed{\to}$ S   :(B−A)/N$\boxed{\to}$ D   :0$\boxed{\to}$ I   :$\boxed{\text{Lbl}}$ 1
:I+1$\boxed{\to}$ I   :A+(I−1+R)*D$\boxed{\to}$ X   :$\boxed{\text{Y}_1}$ +S$\boxed{\to}$ S   :$\boxed{\text{If}}$ I<N
:$\boxed{\text{Goto}}$ 1   :S*D$\boxed{\to}$ S   :$\boxed{\text{Disp}}$ "RIEM="   :$\boxed{\text{Disp}}$ S

Program Step	Explanation
:RIEM	Name the program.
:0$\boxed{\to}$ S :(B−A)/N$\boxed{\to}$ D	Start the sum at S=0. Compute Δx and store the value in the variable D.
:0$\boxed{\to}$ I	Start a loop using counter I.
:$\boxed{\text{Lbl}}$ 1	
:I+1$\boxed{\to}$ I	
:A+(I−1+R)*D$\boxed{\to}$ X	Compute the x-value to plug into f.
:$\boxed{\text{Y}_1}$ +S$\boxed{\to}$ S	Add $f(c_i)$ to the sum.
:$\boxed{\text{If}}$ I<N	Continue the loop if not finished.
:$\boxed{\text{Goto}}$ 1	
:S*D$\boxed{\to}$ S	Multiply the sum by D=Δx.
:$\boxed{\text{Disp}}$ "RIEM=" :$\boxed{\text{Disp}}$ S	Display the Riemann sum.

To use program RIEM, you must first store the values of the left and right endpoints of the interval $[a, b]$, in the variables A and B, respectively, the number of subintervals, n, in the variable N and a value for the parameter R. You must also enter the function f as Y_1.

Users of the TI-85 should store the following, slightly shorter, program under the name RIEM. Note that to get $\boxed{\text{For}}$ you should press $\boxed{\text{CTL}}$ $\boxed{\text{For}}$. To get the $\boxed{\text{End}}$ command, press $\boxed{\text{CTL}}$ $\boxed{\text{End}}$. Also, pay particular attention to the arguments (the symbols inside the parentheses) of the For command: do not confuse the I and the 1 (one).

:0 $\boxed{\rightarrow}$ S   :(B−A)/N $\boxed{\rightarrow}$ D   : $\boxed{\text{For}}$ (I,1,N,1)   :A+(I−1+R)*D $\boxed{\rightarrow}$ x

:S+y1 $\boxed{\rightarrow}$ S   : $\boxed{\text{End}}$:S*D $\boxed{\rightarrow}$ S   : $\boxed{\text{Disp}}$ "RIEM=",S

Example 2. Computing Riemann Sums with RIEM

Compute $R_8(f)$, $R_{25}(f)$, $R_{100}(f)$ and $R_{500}(f)$ for $f(x) = x^2$ on the interval $[0, 1]$ using left-hand ($R = 0$), midpoint ($R = .5$) and right-hand ($R = 1$) evaluations. To run RIEM, we first need to initialize the variables A, B, N, R and Y1. Here, we have

$$0 \boxed{\rightarrow} A \qquad\qquad 1 \boxed{\rightarrow} B$$
$$8 \boxed{\rightarrow} N \qquad\qquad 0 \boxed{\rightarrow} R$$
$$Y1 = x \wedge 2$$

Now, execute program RIEM: on the TI-81, press $\boxed{\text{PRGM}}$ R $\boxed{\text{ENTER}}$ and on the TI-85, press $\boxed{\text{PRGM}}$ $\boxed{\text{NAMES}}$ $\boxed{\text{RIEM}}$ $\boxed{\text{ENTER}}$. You should get RIEM=.2734375. Then change to $R = .5$ (press .5 $\boxed{\rightarrow}$ R) and execute RIEM again. On the TI-85, it is convenient to enter 0.5 $\boxed{\rightarrow}$ R:RIEM on the same line. After the calculator returns RIEM=.33203, press $\boxed{\text{ENTRY}}$ and edit the expression to 1.0 $\boxed{\rightarrow}$ R:RIEM. You should construct the following table of Riemann sums:

N	R=0	R=.5	R=1
8	.2734375	.33203125	.3984375
25	.3136	.3332	.3536
100	.32835	.333325	.33835
500	.332334	.333333	.334334

There are several observations that we can make from these results. First, the sums for $R = .5$ are in between the sums for $R = 0$ and $R = 1$ for each value of N. Note that since $y = x^2$ is an increasing function on $[0, 1]$, we can, in fact, conclude that the left-hand evaluations ($R = 0$) give the smallest function values and, hence, also the smallest Riemann sums. These sums are called *lower sums* . Further, the right-hand ($R = 1$) evaluations for an increasing function f give the largest Riemann sums. These sums are called *upper sums* . Finally, for any other choice of evaluation points, the Riemann sum for a given N will fall in between the corresponding lower and upper sums. Unfortunately, for many functions the lower and upper sums are not of practical value. The maximum and minimum values of the function on each subinterval are not necessarily at the endpoints of the subinterval and are often quite hard to find.

In the table of numbers given above, all three columns appear to be approaching 1/3, as N gets larger. It is possible to show by hand (you can find the details in many calculus books) that both the lower and upper sums approach 1/3. Then, by the Pinching Theorem, *all* Riemann sums approach 1/3 as $n \to \infty$. The following definition should now be meaningful.   ∎

Definition The *definite integral* of the function $f(x)$ over the interval $[a, b]$, denoted by $\displaystyle\int_a^b f(x)\, dx$, is defined by

$$\int_a^b f(x)\, dx = \lim_{n \to \infty} R_n(f)$$

if the limit exists and is the same for every choice of the evaluation points.

This definition may seem to be an abstract mathematical notion, but recall that it was motivated by the more concrete question of how to compute areas.

We started this section by estimating $\displaystyle\int_0^1 (2x - 2x^2)\, dx$, i.e., the area between the parabola and the x-axis seen in Figure 5.1b. You might wonder whether an integral always gives the area of some region. The next example shows that the answer is *no*. While we originally had in mind computing the area under the graph of a function f for which $f(x) > 0$, the definition of the integral will make sense as long as the indicated limit exists. In general, however, the integral can be thought of as representing something called *signed area*. We explore this in the next two examples.

Example 3. Signed Area

Estimate the value of the definite integral $\int_0^1 (x-1)\,dx$ and compare your estimate to the area of the triangle bounded by the lines $x=0$, $y=x-1$ and the x-axis. The desired area is shown in Figure 5.6. This figure may be obtained by graphing $y=x-1$ (we have used the Range parameters Xmin$=-3$, Xmax$=4$, Ymin$=-3$ and Ymax$=2$) and using the command Shade(X$-1,0,1,0,1$) on the TI-81 [press [DRAW] 7(Shade) X$-1,0,1,0,1$) [ENTER]] or Shade(x$-1,0,0,1$) on the TI-85 [with the graph displayed, press [MORE] [DRAW] [Shade] x$-1,0,0,1$[)] [ENTER]].

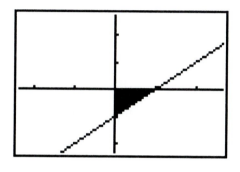

FIGURE 5.6

The TI-85 has a special capability for computing integrals. We will estimate the value of the integral using the program **RIEM** on the TI-81 and the built-in integration routine on the TI-85. Using **RIEM** with R$=.5$, you should get the following sums.

N	RIEM
8	$-.5$
32	$-.5$
128	$-.5$

On the TI-85, press [CALC] [fnInt] x$-1,x,0,1$[)] [ENTER] . The calculator quickly returns $-.5$, which is its estimate of the integral to within a factor of 10^{-5} (the value of the error tolerance variable *tol*, found in the Toler menu).

Based on this evidence, we would feel comfortable with an estimate of $-.5$ for the integral (this is the exact value, as we will see in the next section). Notice that the area of the triangle in Figure 5.6 is $+.5$. The absolute value of the integral gives the correct area. Here, the minus sign indicates that the area lies *below* the x-axis. This is an example of what we mean by *signed area*.   ∎

NOTE: The TI-85 also has a special integral command available from the graphing menu. To use this for Example 3, start by graphing $y = x - 1$ using the range parameters of Figure 5.1a. Press MORE MATH to bring up the Math graphing menu. We have previously used this menu to solve for roots, maxima and minima. This time, press $\int f(x)$. The TRACE cursor will appear at the point $(0, -1)$. Since $x = 0$ is the lower limit of integration, press ENTER . Now, move the cursor to the point $(1,0)$. Since $x = 1$ is the upper limit of integration, press ENTER . At the bottom of the display, you should see "$\int f(x) = -.5$." Note that using the standard range values, we would not have been able to move the cursor to *exactly* $x = 1$. Thus, to use the integral command, we may need to adjust the range parameters so that we can get the correct limits of integration. In this case, the zoom command ZDECM will set the range values to place pixels at $x = .1, .2, .3$, etc., so that we get $x = 1$ exactly. However, we note that if a limit of integration is irrational (e.g., $\sqrt{2}$), it will be *impossible* to get the exact value for the limit, and this imprecision will affect the accuracy of the calculator's answer.

Example 4. Sums of Signed Areas

Estimate $\int_0^1 (x^2 - x)\,dx$, $\int_1^2 (x^2 - x)\,dx$ and $\int_0^2 (x^2 - x)\,dx$ and interpret the integrals as signed areas. We will use RIEM to get estimates of the integrals on the TI-81, and the fnInt command to obtain estimates of the integrals on the TI-85. The following table of numbers comes from RIEM with R=.5:

N	$\int_0^1 (x^2 - x)\,dx$	$\int_1^2 (x^2 - x)\,dx$	$\int_0^2 (x^2 - x)\,dx$
8	−.16796875	.83203125	.65625
32	−.1667480469	.8332519531	.666015625
128	−.1666717529	.8333282471	.6666259766

On the TI-85, press CALC fnInt x∧2−x , x , 0 , 1) ENTER , then press ENTRY and change 0,1 to 1,2. Finally, press ENTRY and change 1,2 to 0,2. We obtain the estimates:

$$\int_0^1 (x^2 - x)\,dx \approx -.1666$$

$$\int_1^2 (x^2 - x)\,dx \approx .8333$$

$$\int_0^2 (x^2 - x)\,dx \approx .6666$$

You might suspect that the exact values of the integrals are $-1/6$, $5/6$ and $2/3$, respectively. (It can be shown that these are in fact, correct. However, you should be careful not to jump to conclusions, as many integrals do not have rational values.) A quick sketch of the graph (see Figure 5.7 for the TI-85 graph) shows that $(x^2 - x) < 0$ on $(0,1)$ so that the area from $x = 0$ to $x = 1$ is $1/6$. Since $(x^2 - x) > 0$ on $(1,2)$, the area from $x = 1$ to $x = 2$ is $5/6$. Now, the total area from $x = 0$ to $x = 2$ in Figure 5.7 is $1/6 + 5/6 = 1$. However,

$$\int_0^2 (x^2 - x)\,dx = -1/6 + 5/6 = 2/3$$

The integral adds up the signed areas, so that the proper interpretation of an integral requires a knowledge of where the function is positive and where it is negative. ∎

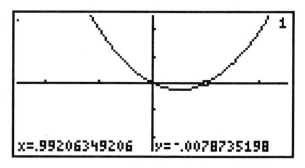

FIGURE 5.7

Note that Example 4 illustrates a general property of integrals, namely that for any c in $[a, b]$,

$$\int_a^b f(x)\,dx = \int_a^c f(x)\,dx + \int_c^b f(x)\,dx$$

We will address one last fundamental question in this section. When does a definite integral exist (i.e., when does the limit defining an integral exist)? This turns out to be an important issue, because although a limited number of integrals can be computed exactly (we will see how to do this in the next section) *most* integrals cannot be computed exactly. We are usually forced to approximate the values of integrals using Riemann sums or some other computational method such as those discussed in section 5.2 (or using the routine built into the TI-85). The problem with this approach is that, unless we have some way of knowing that a given integral exists, we will not know if our numerical approximation has any meaning (since numerical methods for almost any integral will produce *some* number, whether or not the integral actually exists). The question, then, is whether or not the "approximation" approximates anything meaningful. The following result gives a partial answer to this question.

Theorem 5.1 If f is continuous on $[a, b]$ then $\int_a^b f(x)\, dx$ exists.

Notice that the theorem says nothing about what the possible effect of discontinuities might be. The bottom line is: for continuous functions, there are no problems with the existence of the integral. For discontinuous functions, we need to proceed with caution. Example 5 shows how we can use program RIEM to make reasonable conjectures about the existence of such so-called *improper integrals*.

Example 5. Integrals of Discontinuous Functions

Investigate whether or not $\int_0^1 \frac{1}{x}\, dx$ and $\int_0^1 \frac{1}{\sqrt{x}}\, dx$ exist. Since both integrands are discontinuous at $x = 0$, we do not know from Theorem 5.1 whether or not either integral exists. Our program RIEM can help us distinguish between the two cases. Note that in this case we must be careful about our choice of R: R=0 will produce a division by 0. (Why is this?) Using R=.5, we get the following table of Riemann sums:

N	$f(x) = 1/x$	$f(x) = 1/\sqrt{x}$
16	4.736261	1.848856
64	6.122403	1.924392
256	7.508688	1.962194
1024	8.894981	1.981096

This gives us evidence that $\int_0^1 \frac{1}{x} \, dx$ may not exist (since the sums do not seem to be approaching a limit) and that $\int_0^1 \frac{1}{\sqrt{x}} \, dx$ *does* exist (and equals approximately 2). Both of these conjectures, it can be shown, are correct.   ∎

We close with a question that is probably as much philosophical as it is mathematical. Given that $\int_0^1 \frac{1}{\sqrt{x}} \, dx = 2$, is it reasonable to say that 2 is the *area* bounded by the graphs of $x = 0$, $y = 0$ and $y = \frac{1}{\sqrt{x}}$? (Draw the picture for yourself and think about this carefully. HINT: Can you draw the entire graph?)

Exercises 5.1

In exercises 1-2, count pixels to estimate the integrals (use the Range values given for Figure 5.1a).

1. $\int_0^1 x^2 \, dx$

2. $\int_0^2 \sin x \, dx$

In exercises 3-6, use RIEM to compute lower sums and upper sums to show that they approach a common limit.

3. $\int_0^1 x^2 \, dx$

4. $\int_0^1 x^3 \, dx$

5. $\int_0^1 (1 - x^3) \, dx$

6. $\int_0^2 \sqrt{x} \, dx$

In exercises 7-10, the values of the integrals are integers. Use RIEM to discover these values.

7. $\int_0^2 (4x^3 - 7x) \, dx$

8. $\int_0^2 (2x^3 - 3x^2) \, dx$

9. $\int_1^2 3(x - 1)^2 \, dx$

10. $\int_0^8 3\sqrt{x + 1} \, dx$

In exercises 11-18, estimate the areas of the indicated regions. Recall that the integral computes signed areas.

11. The region bounded by $y = x^4 - 1$ and $y = 0$.

12. The region bounded by $y = x^4 - 1$, $y = 0$, $x = 1$ and $x = 2$.

13. The region bounded by $y = x^2$, $y = 0$ and $x = 2$.

14. The region bounded by $y = x^2 - 2x$ and $y = 0$.

15. The region bounded by $y = \sqrt{x}$, $y = 0$ and $x = 3$.

16. The region bounded by $y = x^2$, $y = 0$, $x = -1$ and $x = 2$.

17. The region bounded by $y = \sin x$, $y = 0$, $x = 0$ and $x = 2\pi$.

18. The region bounded by $y = x^3$, $y = 0$, $x = -1$ and $x = 1$.

In exercises 19-22, determine whether or not the numerical evidence suggests that the integral exists.

19. $\int_0^1 \frac{1}{x^2}\, dx$ 20. $\int_0^1 x^{-2/3}\, dx$

21. $\int_2^4 \frac{1}{x-1}$ 22. $\int_1^2 \frac{1}{x-1}\, dx$

23. Argue that $\int_0^2 f(x)\, dx = \int_0^2 g(x)\, dx$ in the case where $f(x) = \begin{cases} 2x, & x \leq 1 \\ 3x^2, & x > 1 \end{cases}$

and $g(x) = \begin{cases} 2x, & x < 1 \\ 3x^2, & x \geq 1 \end{cases}$. HINT: what is the difference between f and g?

24. Estimate the integral in exercise 23.

25. Estimate $\int_0^2 f(x)\, dx$ where $f(x) = \begin{cases} x-1, & x < 1 \\ 4 - x^2, & x \geq 1 \end{cases}$.

26. Investigate whether or not $\int_0^1 \sin\left(\frac{1}{x}\right)\, dx$ exists.

27. The Mean Value Theorem states that if F is differentiable on $[a, b]$ then there exists a number c in (a, b) such that $F'(c)[b - a] = F(b) - F(a)$. Determine c for $F(x) = x^3/3$ on the intervals $[0,1]$, $[1,2]$ and $[2,3]$. Using these c's as evaluation points, show that $R_3(f)$ equals the exact integral $\int_0^3 x^2\, dx = 9$. With the correct choice of c, then, it is possible for $R_n(f)$ to exactly equal the integral for any n.

28. For the integral $\int_0^3 x^2\, dx = 9$, show that there is a value of R (N=3) for which program RIEM gives the exact integral. Carefully state the "mean value theorem" that this result illustrates. How does R compare to the c's found in exercise 27?

29. The following will make RIEM easier for you to use. Under the name RINP (we recommend PrgmQ on the TI-81) store the program:

:Disp "A"   :Input A   :Disp "B"   :Input B
:Disp "N"   :Input N   :Disp "R"   :Input R

Store the function F in Y1 and then run RINP before using RIEM.

EXPLORATORY EXERCISE

Introduction

We have made several references in this section to a method of computing integrals exactly. In this exercise, you will discover this method. The key is to look for a simple rule, so we will assume that $\int_0^c x^n\, dx = a c^b$ for some constants a and b which we will try to determine.

Problems

Compute $\int_0^1 x\, dx$ and $\int_0^2 x\, dx$ and determine a and b to match the conjecture $\int_0^c x\, dx = a c^b$ for $c = 1$ and $c = 2$. That is, estimate $\int_0^1 x\, dx$ and set the estimate equal to $a(1)^b$. Then, estimate $\int_0^2 x\, dx$ and set the estimate equal to $a(2)^b$. Use these two equations to solve for a and b. Now try $\int_0^3 x\, dx$ and $\int_0^4 x\, dx$ and see if the formula works for the a and b just found. Repeat the above procedure (that is, find new values for a and b) with $\int_0^c x^2\, dx$ and $\int_0^c x^3\, dx$. Now, look at your solutions.

In the general formula $\int_0^c f(x)\, dx = g(c)$ what is the relationship between f and g? Test your conjecture on $\int_0^2 (x^2 - 2)\, dx$ and $\int_0^{\pi/2} \cos x\, dx$.

Further Study

You have several pieces of what is known as the Fundamental Theorem of Calculus. It only remains to extend your conjecture from this exercise to $\int_a^b f(x)\, dx$ for $a \neq 0$ and to prove the result (the standard proof uses the ideas of exercise 27).

5.2 Computation of Integrals

In section 5.1, we introduced the notion of the definite integral, defining it as a limit of Riemann sums. While this definition is theoretically quite important and provides us with a very straightforward definition of area, we must point out that integrals are in practice only rarely computed (or approximated) using Riemann sums. In this section, we look at more sophisticated techniques for approximating integrals, as well as a technique which will compute a limited number of integrals exactly.

THE FUNDAMENTAL THEOREM OF CALCULUS

The relationship between the integral and the derivative is a remarkable fact which brings unity to the seemingly disjoint studies of differential and integral calculus. You were asked to discover this relationship in an exploratory exercise in the previous section, and you can find a careful proof of the result in your regular calculus book. Its great significance is underscored by its name.

Theorem 5.2 (The Fundamental Theorem of Calculus) Suppose that f is continuous on the interval $[a, b]$. Let $F(x)$ be any function satisfying $F'(x) = f(x)$ for all x in $[a, b]$. (F is then called an *antiderivative* of f.) Then

$$\int_a^b f(x)\, dx = F(b) - F(a)$$

To evaluate $\int_a^b f(x)\, dx$, then, we need only find an antiderivative of f (any one at all will do) and plug in the limits of integration (b and a).

This is a vitally important result because it gives us *exact* answers, and in some cases it is very easy to implement. Unfortunately, in many other cases it is quite difficult to find an antiderivative. Worse yet, many functions do not have antiderivatives which can be written in terms of the elementary functions with which we are all familiar. While the TI-81/85 will not help you find any antiderivatives, the calculator can be very helpful in evaluating the antiderivative at the endpoints.

Example 1. Exact Integrals on the TI-81/85

Compute the area of the portion of $y = -x^2 + x + 5$ above the x-axis. We start with a sketch of the region (see Figure 5.8 for a TI-81 graph). The desired area is $\int_a^b (-x^2 + x + 5)\, dx$, where a and b are the zeros of $-x^2 + x + 5$ (why is this?).

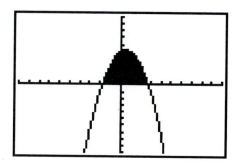

FIGURE 5.8

We compute an antiderivative of $f(x) = -x^2 + x + 5$ and get $F(x) = -x^3/3 + x^2/2 + 5x$. The Fundamental Theorem tells us that the area is $F(b) - F(a)$, but we still need to determine b and a. Using the quadratic formula, we find $a = \dfrac{-1 + \sqrt{21}}{-2} = \dfrac{1 - \sqrt{21}}{2}$ and $b = \dfrac{-1 - \sqrt{21}}{-2} = \dfrac{1 + \sqrt{21}}{2}$. We could, of course, plug these expressions into F, but even if we could simplify the expressions correctly (Try it! You should get $7\sqrt{21}/2$.) we would gain little feeling for the size of the area.

On the TI-81/85, compute $b \approx 2.791287847$ and plug into F using the function evaluation program FEVAL (when prompted for x, enter Ans): $F(b) \approx 10.6028408$. Then compute $a \approx -1.791287847$ and plug into F: $F(a) \approx -5.436174133$. Subtracting, we get

$$\int_a^b -x^2 + x + 5 \, dx \approx 10.6028408 + 5.436174133 = 16.03901493$$

The area is slightly over 16. ∎

Note that for more complicated functions, we might need to use a rootfinding method (see Chapter 4) to get approximations for the endpoints a and b. While owning a TI-81/85 is useful in this way, it is no substitute for learning the standard techniques of integration. (These are discussed in your regular calculus text.)

TRAPEZOID RULE

Due to our inability to find antiderivatives for many functions [try, for example, to find an antiderivative for $\sin(x^2)$, but don't try for too long] we need to supplement the Fundamental Theorem with effective approximation methods. We have already seen one such method, commonly called the *midpoint rule* (although we have not used that name before). The midpoint rule is a Riemann sum with evaluation points equal to the midpoints of the various subintervals (i.e., program **RIEM** with $R = .5$). As was the case for the rootfinding methods discussed in Chapter 4, it is important for us to have several methods available, so that we can balance accuracy and simplicity for a wide range of problems.

The *trapezoid rule*, which we describe below, is in many ways similar to the midpoint rule. Along with having a nice geometric interpretation, the trapezoid rule is significant because of its extension to a more powerful rule called *Simpson's rule*, to be discussed shortly.

We first note that, in practice, we may not know the function which we're trying to integrate. That's right: we often will know only some *values* of a function at a collection of points. An algebraic representation of the function might not be available. For example, in experiments in the physical and biological sciences, it is usually the case that the only information available about a function comes from measurements made at a finite number of points.

Example 2. Estimating Area from Data with Trapezoids

Use the function values given in the following table to estimate the area bounded by the graphs of $x = 0$, $x = 1$, and the (unknown) function which generated the data.

x:	0	0.25	0.5	0.75	1.0
$f(x)$:	1	1.3	1.8	1.6	1.6

Conceptually, we have two tasks: first to conjecture a reasonable way to connect the given points, and then to estimate the area. The simplest way to connect the dots is with straight-line segments (at least, this is the way that the eight-year-old daughter of one of the authors does it; see Figure 5.9). Notice that the region thus constructed is composed of 4 trapezoids. Recall that the area of a trapezoid with sides h_1 and h_2 and base b is given by $b*(h_1+h_2)/2$. This is the sum of the area of a rectangle plus the area of a triangle. (Why?) The total area is then

$$.25\frac{f(0)+f(.25)}{2}+.25\frac{f(.25)+f(.5)}{2}+.25\frac{f(.5)+f(.75)}{2}+.25\frac{f(.75)+f(1)}{2}$$

$$=[f(0)+2f(.25)+2f(.5)+2f(.75)+f(1)]\frac{.25}{2}=1.425$$

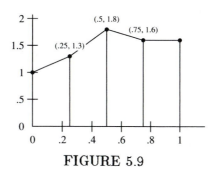

FIGURE 5.9

The trapezoid rule is a generalization of our work in Example 2. We first divide the interval $[a,b]$ into n equal pieces with endpoints $a=x_0<x_1<x_2<...<x_n=b$. The $(n+1)$-point trapezoid rule approximation of $\int_a^b f(x)\,dx$ is then:

$$T_n(f)=[f(x_0)+2f(x_1)+2f(x_2)+...+2f(x_{n-1})+f(x_n)]\frac{b-a}{2n}$$

In the exercises, you are asked to show that $T_n(f)=(A_n+B_n)/2$, where A_n is the Riemann sum from program RIEM with $R=0$ and B_n is the Riemann sum with $R=1$. This suggests the following simple (although not particularly efficient) program for implementing the trapezoid rule. The first program given is for the TI-81. We suggest that TI-81 users store this program as PrgmT. Note that to enter the command $\boxed{\text{PrgmR}}$ you press $\boxed{\text{PRGM}}$ $\boxed{<}$ R.

:TRAP   :0$\boxed{\rightarrow}$ R   :$\boxed{\text{PrgmR}}$:S$\boxed{\rightarrow}$ T   :1$\boxed{\rightarrow}$ R   :$\boxed{\text{PrgmR}}$
:(S+T)/2$\boxed{\rightarrow}$ T   :$\boxed{\text{Disp}}$ "TRAP="   :$\boxed{\text{Disp}}$ T

Program Step	Explanation
:TRAP	Name the program.
:0 $\boxed{\rightarrow}$ R	Compute A_n and store it in T.
:$\boxed{\texttt{PrgmR}}$	
:S $\boxed{\rightarrow}$ T	
:1 $\boxed{\rightarrow}$ R	Compute B_n.
:$\boxed{\texttt{PrgmR}}$	
:(S+T)/2 $\boxed{\rightarrow}$ T	Compute $(A_n + B_n)/2$.
:$\boxed{\texttt{Disp}}$ "TRAP="	Display the trapezoid approximation.
:$\boxed{\texttt{Disp}}$ T	

The TI-85 program follows. When prompted for a name, enter TRAP.

:0 $\boxed{\rightarrow}$ R   :RIEM   :S $\boxed{\rightarrow}$ T   :1 $\boxed{\rightarrow}$ R   :RIEM
:(S+T)/2 $\boxed{\rightarrow}$ T   :$\boxed{\texttt{Disp}}$ "TRAP=",T

Example 3. Comparison of the Midpoint and Trapezoid Rules

For $\int_0^1 3x^2\,dx$, compare the midpoint and trapezoid rules with $n = 4$, $n = 8$ and $n = 16$ to the exact value of the integral. For the midpoint rule, we use RIEM with $R = .5$ and we use the new program TRAP for the trapezoid rule. Both programs require A, B, N and Y1 to be initialized. Here, we have

0 $\boxed{\rightarrow}$ A   1 $\boxed{\rightarrow}$ B   4 $\boxed{\rightarrow}$ N   Y1=3X$\wedge$2

Execute PrgmR on the TI-81 (or RIEM on the TI-85) and PrgmT on the TI-81 (or TRAP on the TI-85). Note that after executing TRAP, the current value of R will be 1. You will need to change R back to .5 to get the next midpoint approximation. Change N to 8 and execute both programs again. Change N to 16 and execute both programs. We get the following table of values.

N	Midpoint	Trapezoid
4	.984375	1.03125
8	.99609375	1.0078125
16	.99902343	1.001953125

Of course, from the Fundamental Theorem,

$$\int_0^1 3x^2 \, dx = \left[\, x^3 \, \right]_0^1 = 1$$

exactly. The errors for the two rules here are fairly close, although the midpoint rule is slightly more accurate. This accuracy comparison is typical. The geometry of the two rules and the fact that $y = 3x^2$ is concave up on $[0,1]$ should explain why the midpoint rule gives values that are too low and the trapezoid rule gives ones that are too high. (Draw a picture and think about this carefully.)   ∎

EXTRAPOLATION

Mathematicians are in the business of trying to recognize, explain and make use of patterns. The approximations in Example 3 will reveal a pattern which we can take advantage of to obtain a better approximation.

Let's look more closely at the trapezoid rule approximations. Note that $T_n(f)$ is getting smaller as n gets larger. From this pattern, we would expect the integral to be smaller than 1.001953125. In fact, we can be precise about how much smaller we expect the answer to be. Note that $T_4 - T_8 = .0234375$ and $T_8 - 1 = .0078125$. (Recall that the exact value of the integral is 1.) Also, $T_8 - T_{16} = .005859375$ and $T_{16} - 1 = .001953125$. Now for a surprise:

$$\frac{T_4 - T_8}{T_8 - 1} = \frac{T_8 - T_{16}}{T_{16} - 1} = 3.0$$

Both ratios are *exactly* equal to 3! In other words, if we represent the exact integral by I, we get

$$T_4 - T_8 = 3 * (T_8 - I) \quad \text{and} \quad T_8 - T_{16} = 3 * (T_{16} - I)$$

Solving this for I, we get $3I = 3T_8 + (T_8 - T_4)$ and $3I = 3T_{16} + (T_{16} - T_8)$. Finally, this leaves us with

$$I = T_8 + (T_8 - T_4)/3 \quad \text{and} \quad I = T_{16} + (T_{16} - T_8)/3$$

It turns out that the following result is true, in general: T_8 is about $3/4$ of the way from T_4 to the exact integral I, and T_{16} is about $3/4$ of the way from T_8 to the exact integral. That is,

$$I \approx T_8 + (T_8 - T_4)/3 \quad \text{and} \quad I \approx T_{16} + (T_{16} - T_8)/3$$

That's a pattern we can take advantage of! We make the following definition.

Definition The *Richardson extrapolation* of the trapezoid rule is given by $E_{2n} = T_{2n} + (T_{2n} - T_n)/3$.

For Example 3, $E_8 = T_8 + (T_8 - T_4)/3 = 1$ and $E_{16} = T_{16} + (T_{16} - T_8)/3 = 1$. Thus, in this example we have taken two trapezoid approximations and the Richardson extrapolation has led us directly to the exact value of the integral! In general, the extrapolation will not give the exact value of an integral, but it will greatly increase the accuracy of our approximation (usually much more so than by simply increasing n alone).

SIMPSON'S RULE

It turns out that the extrapolation of the trapezoid rule has a very simple form. Returning to Example 3, some messy but basic algebra gives us

$$E_8 = \frac{f(0)}{24} + 4\frac{f(.125)}{24} + 2\frac{f(.25)}{24} + 4\frac{f(.375)}{24} + 2\frac{f(.5)}{24}$$
$$+ 4\frac{f(.625)}{24} + 2\frac{f(.75)}{24} + 4\frac{f(.875)}{24} + \frac{f(1)}{24}$$

Notice the common factor of $1/24$ and the pattern that the coefficients follow: 1, 4, 2, 4, 2, 4, 2, 4, 1. This is an example of Simpson's rule, which has the general form

$$S_n(f) = [f(x_0) + 4f(x_1) + 2f(x_2) + 4f(x_3) + \ldots + 4f(x_{n-1}) + f(x_n)]\frac{b-a}{3n}$$

where we have used the same notation as for the trapezoid rule. You should note here that the value of n must be even. (Why is that?)

Since Simpson's rule is an extrapolation of the trapezoid rule, it is, in general, much more accurate than either the trapezoid rule or the midpoint rule. In addition,

Simpson's rule has a nice geometric interpretation. Recall that in Example 2 we connected 2 points at a time with line segments to form trapezoids. If we, instead, connect 3 points at a time with parabolas, we get Simpson's rule (although the algebra involved in showing this is quite messy). Since parabolas would give a smoother, perhaps more reasonable looking graph, we have a geometric explanation of the increased accuracy of Simpson's rule.

The following program is a simple (although, like program TRAP, not especially efficient) program for computing Simpson's rule approximations. It uses the program TRAP and requires you to initialize the same variables as TRAP does. We suggest that TI-81 users enter this as PrgmS.

:SIMP   :[PrgmT]   :T[→] P   :N/2[→] N   :[PrgmT]   :P+(P−T)/3[→] P
              :2N[→] N   :[Disp] "SIMP="   :[Disp] P

Program Step	Explanation
:SIMP	Name the program.
:[PrgmT]	Compute T_n and store it in P.
:T[→] P	
:N/2[→] N	Divide N by 2 .
:PrgmT	Compute $T_{n/2}$.
:P+(P−T)/3[→] P	Compute E_n.
:2N[→] N	Restore the value of N.
:[Disp] "SIMP="	Display the Simpson approximation.
:[Disp] P	

The TI-85 program follows. When prompted for a name, enter SIMP.

:TRAP   :T[→] P   :N/2[→] N   :TRAP
:P+(P−T)/3[→] P   :2N[→] N   :[Disp] "SIMP=",P

For both of these programs, you will need to store values for A, B, N (which must be even) and Y1.

Example 4. Simpson's Rule Approximations

Use Simpson's rule to estimate $\int_0^2 \sqrt{x^2 + 1} \, dx$. We will not reference an exact value, but rather try to decide what seems reasonable. To use SIMP, we must initialize A, B, N and Y1. In the present case, enter

$$0 \boxed{\rightarrow} \text{A} \quad 2 \boxed{\rightarrow} \text{B} \quad 4 \boxed{\rightarrow} \text{N}$$
$$\text{Y1} = \sqrt{} (\text{X} \wedge 2 + 1)$$

Executing SIMP repeatedly for different values of N (we have used powers of 2), we generate the following approximations.

N	Simpson's Rule
4	2.957955601
8	2.957883497
16	2.95788557

Since we expect Simpson's rule to be very accurate, the agreement of the first 6 digits of S_8 and S_{16} leads us to conjecture that 2.95788 is a good approximation.

■

With Simpson's rule, you have a generally very accurate numerical integration method. Many calculus books include formulas for the errors made by the trapezoid rule and Simpson's rule. These formulas are typically used to give worst-case error estimates, and clearly indicate the superiority of Simpson's rule. You will discover the relative accuracies of the trapezoid and Simpson's rules in the exercises.

More importantly, the derivation of Simpson's rule will have acquainted you with some of the general concepts behind the numerical methods built into most computer integration software.

Exercises 5.2

In exercises 1-6 use the Fundamental Theorem as in Example 1 to find the area between the given function and the x-axis.

1. $5 - x^2$

2. $-x^2 + 3x + 5$

3. $-x^2 - x + 4$

4. $-x^4 + 3$

5. $-x^4 + 3x^2 + 5$

6. $-x^4 + x^3 + 5x + 4$

In exercises 7-10, use the trapezoid rule and Simpson's rule as in Example 2 to approximate $\int_0^1 f(x)\,dx$.

7.

x:	0	0.25	0.5	0.75	1
$f(x)$:	2	2.4	3.0	3.3	3.6

8.

x:	0	0.25	0.5	0.75	1
$f(x)$:	3	2.1	2.7	3.4	4.2

9.

x:	0	0.125	0.25	0.375	0.50	0.625	0.75	0.875	1
$f(x)$:	1	1.3	1.5	1.6	1.6	2.0	2.4	2.9	3.5

10.

x:	0	0.125	0.25	0.375	0.50	0.625	0.75	0.875	1
$f(x)$:	2	1.2	0.4	−.5	0	0.4	1.2	2.5	4.0

In exercises 11-14, compare the midpoint, trapezoid and Simpson's rule approximations for N=4, N=8 and N=16 to the exact value.

11. $\int_0^2 (3x^2 - 1)\,dx$

12. $\int_0^3 (3x^2 - 2x + 1)\,dx$

13. $\int_1^2 (x - 1)^4\,dx$

14. $\int_0^1 \dfrac{4}{x^2 + 1}\,dx \qquad (= \pi)$

In exercises 15-20, estimate the following integrals (4 digits accuracy):

15. $\int_0^1 x\sqrt{x^3 + 1}\,dx$

16. $\int_0^2 \sqrt{x^3 + 1}\,dx$

17. $\int_0^2 1/\sqrt{x^2 + 1}\,dx$

18. $\int_0^2 1/\sqrt{x^3 + 1}\,dx$

19. $\int_0^2 \dfrac{2}{\sqrt{4 - x^2}}\,dx$

20. $\int_0^2 \dfrac{\sin x}{x}\,dx$

21. Show that $T_n = (A_n + B_n)/2$ for $n = 4$ and $n = 8$, as described in the discussion before program TRAP. Also, show that $S_{2n} = (T_n + 2M_n)/3$, where M_n is the midpoint rule.

22. Use SIMP to try to estimate $\int_0^2 1/\cos x\,dx$. Graph $1/\cos x = \sec x$ to help your interpretation of the results.

In exercises 23-26, determine which rules (midpoint, trapezoid, Simpson's) give the exact integral for $n = 4$. Explain your results geometrically.

23. $\displaystyle\int_0^1 4x\,dx$ 24. $\displaystyle\int_0^2 x^2\,dx$

25. $\displaystyle\int_0^2 4x^3\,dx$ 26. $\displaystyle\int_0^1 5x^4\,dx$

27. As you might expect, Simpson's rule can be extrapolated in a similar way to our extrapolation of the trapezoid rule. Use $\displaystyle\int_0^1 5x^4\,dx$ and compute $\dfrac{S_4 - S_8}{S_8 - 1}$ and $\dfrac{S_8 - S_{16}}{S_{16} - 1}$. Derive a formula for E_{2n} in terms of S_{2n} and S_n.

28. Use your extrapolation formula from exercise 27 to improve your estimates from exercise 15 to 8 digits accuracy.

In exercises 29-32, we explore how the TI-81/85 can help with the integration technique of partial fractions. To integrate exercise 29, the integrand must be rewritten in the form

$$\frac{2x + 1}{x^2(x^2 + 4)} = \frac{a}{x} + \frac{b}{x^2} + \frac{cx + d}{x^2 + 4}$$

To solve for b, multiply by x^2 and set $x = 0$: $\frac{1}{4} = b$. Unfortunately, solving for a, c and d is messier. Plug in $x = 1$: $\frac{3}{5} = a + b + \frac{1}{5}c + \frac{1}{5}d$. With $b = \frac{1}{4}$, this reduces to $a + \frac{1}{5}c + \frac{1}{5}d = \frac{7}{20}$. Plug in $x = -1$ and simplify to get $-a - \frac{1}{5}c + \frac{1}{5}d = \frac{-9}{20}$. Plug in $x = 2$ and simplify to get $\frac{1}{2}a + \frac{1}{4}c + \frac{1}{8}d = \frac{3}{32}$. Thus, we have

$$\begin{pmatrix} 1 & 1/5 & 1/5 \\ -1 & -1/5 & 1/5 \\ 1/2 & 1/4 & 1/8 \end{pmatrix} = \begin{pmatrix} 7/20 \\ -9/20 \\ 3/32 \end{pmatrix}$$

To solve this on the TI-81/85 (see the exploratory exercise in section 1.3 or your owner's manual for more complete instructions), you must enter 3 × 3 for the size of the matrix A, enter the elements of A in the correct order (you may enter them in fraction form and the calculator will convert them into decimal), enter 3 × 1 for the size of B, enter the elements of B and compute $A^{-1}B$. After determining a-d, integrate.

Use partial fractions to integrate the following.

29. $\displaystyle\int \frac{2x+1}{x^2(x^2+4)}\,dx$

30. $\displaystyle\int \frac{2x^3-x-1}{x^2(x^2+x+1)}\,dx$

31. $\displaystyle\int \frac{3x+6}{(x^2+1)(x^2+4)}\,dx$

32. $\displaystyle\int \frac{2x^3+x}{(x^2+1)(x^2-x+1)}\,dx$

EXPLORATORY EXERCISE

Introduction

We have seen how simple extrapolation formulas can greatly improve the accuracy of approximations. Specifically, we derived Simpson's rule $S_{2n} = T_{2n} + \dfrac{T_{2n}-T_n}{3}$. In exercise 27, you were asked to derive an extrapolation of Simpson's rule $E_{2n} = S_{2n} + \dfrac{S_{2n}-S_n}{15}$. We will continue to extrapolate our extrapolations in this exercise.

Problems

From the trapezoid calculations T_4, T_8, T_{16}, T_{32} and T_{64} we want to extrapolate as far as possible. We have two levels of extrapolation so far, as illustrated below.

$$T_4 \qquad T_8 \qquad T_{16} \qquad T_{32} \qquad T_{64}$$
$$S_8 \qquad S_{16} \qquad S_{32} \qquad S_{64}$$
$$E_{16} \qquad E_{32} \qquad E_{64}$$

We want to extrapolate the E's to get two improved approximations (call them F_{32} and F_{64}) and then extrapolate to a "best" approximation (call it G_{64}). We will use $\displaystyle\int_0^2 x^{31}\,dx = 134217728$ to guide our thinking. First, fill in the above chart (compute the T's, S's and E's). We will guess the extrapolation formula for the F's. The formulas for the first two extrapolations differ only in a change of constants from 3 to 15. Since both 3 and 15 are 1 less than a power of 2 ($3 = 2^2 - 1$ and $15 = 2^4 - 1$) we look for the best k for the formula $F_{2n} = E_{2n} + \dfrac{E_{2n}-E_n}{2^k-1}$. Which works best, $k = 4$, $k = 5$ or $k = 6$? Determine k and compute F_{32} and F_{64}. Then find the best m for the formula $G_{2n} = F_{2n} + \dfrac{F_{2n}-F_n}{2^m-1}$.

Further Study

We relied on numerical evidence to find a sequence of extrapolation formulas. A good book on numerical analysis (see *Numerical Analysis*, 4th edition, by Burden and Faires) will show the theoretical explanation for the accuracy of these methods. The method derived above is typically called *Romberg integration*.

5.3 Applications of Integration

In the course of developing the notion of the definite integral, we have discovered three interpretations of integration. Our original motivation was the need for computing areas. We then defined the definite integral as a limit of Riemann sums. Finally, the Fundamental Theorem of Calculus related the definite integral to antiderivatives. Each of these three interpretations is important, and we give a sampling of applications below.

We first look at applications based on the area interpretation of the integral.

Example 1. Area between Curves

Find the area between the graphs of $y = \cos x$ and $y = x^2 - 1$. Figure 5.10 shows the TI-81 graph of the region we are interested in.

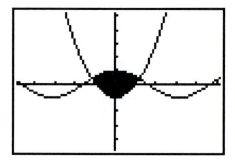

FIGURE 5.10

In this region, notice that $\cos x > x^2 - 1$ so we have

$$\text{Area} = \int_a^b [\cos x - (x^2 - 1)]\, dx$$

where a and b are the x-coordinates of the points of intersection of the two curves. The points of intersection are not easily found, but can be estimated on the calculator using the rootfinding methods of Chapter 4.

First, store Y1=COS(X) and Y2=X²−1, then graph the two functions simultaneously using the standard graphing window (Figure 5.10 has been zoomed in from the standard graph). Using the TRACE command, the closest we can get to the apparent points of intersection is $x \approx -1.157895$ and $x \approx 1.578947$. Now, clear the graph and replace Y1 with COS(X)−(X² − 1) and Y2 with −SIN(X)−2X (the derivative of Y1). Enter -1.157 and execute the Newton's method program NEW-TON (see section 4.1). You should get $x_3 = x_4 = -1.17650194$. To approximate the other root, enter 1.157 and execute NEWTON to obtain $x_3 = x_4 = 1.17650194$.

Users of the TI-85 may instead use the $\boxed{\text{ISECT}}$ command: after graphing $\cos x$ and $x^2 - 1$, press $\boxed{\text{MORE}}$ $\boxed{\text{MATH}}$ $\boxed{\text{MORE}}$ $\boxed{\text{ISECT}}$, move the cursor to the apparent location of an intersection and press $\boxed{\text{ENTER}}$ twice. Then press $\boxed{\text{GRAPH}}$ and repeat the process for the other point of intersection. You should verify that $a \approx -1.1765$ and $b \approx 1.1765$.

Since $F(x) = \sin x - x^3/3 + x$ is an antiderivative of the integrand, we have by the Fundamental Theorem that the area is

$$\sin(b) - b^3/3 + b - \sin(a) + a^3/3 - a \approx 3.114$$

To evaluate the above quantity, we store the antiderivative 'SIN(X)$-$X$\wedge$3/3+X' as Y1 and use the function evaluation program FEVAL to compute $F(b)$ and $F(a)$. ∎

You may already be familiar with the notion of a normal probability distribution. For instance, standardized test scores are often normally distributed. That is, if you draw a graph of the frequency of various scores against the scores themselves, the graph will look roughly bell-shaped. For ease of computation, statisticians often refer to the *standard normal distribution*. [This is the normal distribution where the mean (i.e., numerical average) is 0 and where the standard deviation is 1.] This distribution is described by the graph of

$$y = \frac{1}{\sqrt{2\pi}} e^{-x^2/2}$$

(see Figure 5.11 for a TI-85 graph of this function). The probability that a given score falls between two values a and b is then the area under the graph between $x = a$ and $x = b$. We discover an important property of this distribution in Example 2.

Example 2. An Application to Probability Theory

Given that the scores from a certain test are normally distributed, find the probability that a randomly selected score falls within 1 standard deviation of the mean. The probability is then the area underneath the bell-shaped curve (see Figure 5.11) between $x = -1$ and $x = 1$,

$$\int_{-1}^{1} \frac{1}{\sqrt{2\pi}} e^{-x^2/2}\, dx$$

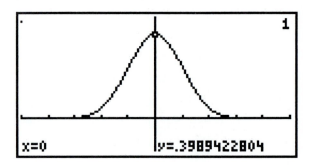

FIGURE 5.11

Unfortunately, it turns out that there is no elementary antiderivative of $e^{-x^2/2}$. (Try to find one, but don't spend much time on it.) Therefore, we estimate the probability using Simpson's rule. We find that $S_8 = .6827$ and $S_{16} = .6827$, and hence conjecture that the probability is about .6827 (i.e., roughly 68.27% of all scores will fall within 1 standard deviation of the mean).   ∎

You have experienced many situations where a force applied to an object changes the velocity of the object. For example, applying the brakes will decrease the velocity of a car. Two factors which determine how much the car slows down are the size of the force (corresponding to how far you depress the brake pedal) and the length of time the force is applied. Physicists use a quantity called *impulse* to measure this effect. For a constant force F applied over a length of time t, the impulse is simply Ft. Newton's 2nd law of motion tells us that $Ft = m\Delta v$ where m is the mass of the object and Δv is the change in velocity of the object. However, forces are rarely constant, so the general form of impulse (which we denote by J) is needed:

$$J = \int_a^b F(t)\, dt$$

Here, a variable force $F(t)$ is applied from time $t = a$ to $t = b$. The general *impulse-momentum* equation is then $J = m\Delta v$.

Example 3. The Impulse of a Baseball

Suppose that a baseball, traveling at 130 ft/sec (about 90 mph), collides with a bat. A sensor on the ball records the force of the bat on the ball every .0001 seconds. The data (taken from *The Physics of Baseball* by R.K. Adair) is given below. What is the speed of the ball after the collision? If we ignore the energy lost due to friction (this aspect is discussed in the exercises), we may use the impulse-momentum equation: $J = m\Delta v$ where m is the mass of the ball (we'll use .01 slugs; regardless of what this name may remind you of, slugs is the correct unit of mass – as opposed to weight – in the English system), Δv is the change of velocity of the ball in feet/sec, and J is the impulse (since J is given by the integral of F, we can interpret the impulse as the area under the graph of the force function). Then, $\Delta v = 100J$ and we can use Simpson's rule on the data to estimate the area J.

F (lb):	0	1250	4250	7500	9000	5500	1250	0	0
t (sec):	0	.0001	.0002	.0003	.0004	.0005	.0006	.0007	.0008

We have $n = 8$ (i.e., there are 9 data points) and

$$S_8 = [0 + 4(1250) + 2(4250) + 4(7500) + 2(9000)$$

$$+ 4(5500) + 2(1250) + 4(0) + 0]\frac{.0008}{24} \approx 2.866$$

Our estimate of the change in velocity is $100(2.866) = 286.6$ ft/sec. The ball then exits the collision traveling in the opposite direction at approximately $286.6 - 130 = 156.6$ ft/sec (107 mph). ∎

In examples 4 and 5, we use integrals to compute geometric quantities other than area (a standard measure of the size of a 2-dimensional region). For a curve in 2 dimensions, the basic measure is length, and for 3-dimensional solids, one of the basic measures is volume. Formulas for the length of a curve and the volume of a solid are derived in your regular calculus book. The format of these derivations is:

1. Divide the curve or solid into a number of pieces.
2. Approximate the measure of each piece and add the approximations.
3. Take the limit of the approximations as the number of pieces tends to infinity.

Does this sound familiar? Such a limit of sums will typically be represented as an integral, as we did in the definition of definite integral in section 5.1.

Example 4. Arclength of a Curve

Find the length of that portion of the parabola $y = -\dfrac{x^2}{500} + \dfrac{x}{2}$ which lies above the x-axis. The x-intercepts occur where $y = 0$, here at $x = 0$ and $x = 250$. This parabola could represent the path of a thrown ball. We would normally say the throw is 250 feet long, which is the horizontal distance covered. The actual distance traveled by the ball is given by the arclength formula

$$s = \int_0^{250} \sqrt{1 + [f'(x)]^2}\, dx = \int_0^{250} \sqrt{1 + (-x/250 + 1/2)^2}\, dx$$

We estimate the integral using Simpson's rule on the TI-81/85 or the $\boxed{\text{fnInt}}$ command on the TI-85, obtaining a distance of about 260 feet. We should emphasize that the arclength formula only rarely produces an integral which can be computed exactly (by finding an antiderivative). The TI-81/85 can thus be invaluable in computing these integrals.   ∎

Example 5. Volume of a Solid of Revolution

Find the volume of the solid obtained by revolving the region bounded by the graphs of $y = \sin x$ and $y = (x - 1)^2$ about the x-axis. As in Example 1, we first graph the region (see Figure 5.12 for the TI-81 graph) by simultaneously graphing Y1=$\sin x$ and Y2=$(x - 1)^2$, and then finding the approximate points of intersection of the graphs.

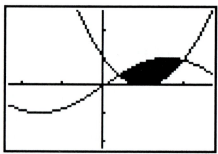

FIGURE 5.12

We use Newton's method (or another rootfinding method) on the TI-81/85 or $\boxed{\text{ISECT}}$ on the TI-85 (see Example 1) to find that the x-coordinates of the points of intersection are approximately $a = .386237$ and $b = 1.961569$. The volume is then given by

$$\text{Volume} = \pi \int_a^b [(f(x))^2 - (g(x))^2]\, dx$$

$$= \pi \int_a^b [(\sin x)^2 - (x - 1)^4]\, dx$$

where we have used the *method of washers* to set up the integral. We now have several options. It is possible to find an antiderivative and evaluate the integral exactly. (Try this yourself to get the exact value.) On the calculator, the integral can be approximated as before, using SIMP or $\boxed{\text{fnInt}}$. The volume is then found to be approximately $.956\pi$. ∎

Examples 6 and 7 utilize the relationship between the definite integral and the antiderivative. Since the derivative is the instantaneous rate of change of a quantity, the integral lets us work from a knowledge of the rate of change of a quantity back to a knowledge of the quantity itself.

Example 6. Computing Distance from Velocity

A runner moves with velocity $v(t) = \dfrac{36t}{\sqrt{t^2 + 2}}$ ft/sec t seconds into a race. How far does she run in the first 10 seconds? Since velocity is the derivative of distance, we want $\int_0^{10} v(t)\, dt$. You can find an antiderivative of $\dfrac{36t}{\sqrt{t^2 + 2}}$ (hint: substitute $u = t^2 + 2$) and find the exact answer of $36(\sqrt{102} - \sqrt{2})$, which you can estimate on your calculator as 312.67 feet. Alternatively, we can use SIMP or $\boxed{\text{fnInt}}$ on the TI-85 to approximate the integral $\int_0^{10} \dfrac{36x}{\sqrt{x^2 + 2}}\, dx$. ∎

Example 7. Computing the Volume of Oil Flow

Suppose that the rate of flow of oil through a pipe is given by $f(t) = t(2 + \sin t)/(3 + t)$ gallons per minute at time t (minutes). Find the amount of oil passing through the pipe in the first 15 minutes. If $A(t)$ is the number of gallons passing through the pipe in the first t minutes, then $A'(t) = f(t)$. Integrate both sides of this equation from $t = 0$ to $t = 15$:

$$\int_0^{15} A'(t)\, dt = \int_0^{15} f(t)\, dt$$

The Fundamental Theorem applies to the left side of this equation to yield

$$A(15) - A(0) = \int_0^{15} f(t)\, dt$$

Finally, since $A(0) = 0$, we have $A(15) = \int_0^{15} f(t)\,dt$. To estimate the integral on the TI-81/85, we need to change the variable from t to x. Using SIMP on the TI-81/85 or $\boxed{\text{fnInt}}$ on the TI-85, we get $A(15) \approx 20$ gallons.   ∎

Exercises 5.3

In exercises 1-4, find the area between the curves.

1. $y = x^4$ and $y = 1 - x$ 2. $y = x^4$ and $y = \cos x$

3. $y = 2x^2 - 1$ and $y = x + 1$ 4. $y = \sin x$ and $y = -x^2$

In exercises 5-8, find the indicated probabilities.

5. A sample of a normal random variable is within 2 standard deviations of the mean (see Example 2).

6. Repeat exercise 5 for 3 standard deviations.

7. The lifetimes of some products are exponentially distributed. Compute the probability that a light bulb lasts less than 20 hours if the probability is given by $\int_0^{20} \frac{1}{10} e^{-x/10}\,dx$.

8. The probability that an electron is between a and b meters from its nucleus is modeled by $\frac{4}{a_0^3} \int_a^b r^2 e^{-2r/a_0}\,dr$ where a_0 is the Bohr radius, 5.29×10^{-11}m. Find this probability for $a = .5a_0$ and $b = a_0$.

In exercises 9-10, repeat Example 3 for the given data (which represent impact velocities of 89 mph and 58 mph, respectively).

9.

t:	0	.0001	.0002	.0003	.0004	.0005	.0006	.0007	.0008
F:	0	1000	2100	4000	5000	5200	2500	1000	0

10.

t:	0	.0001	.0002	.0003	.0004	.0005	.0006	.0007	.0008
F:	0	600	1200	2000	2500	3000	2500	1100	300

In exercises 11-12 the data represent a landowner's measurements in feet of the depth of a lot at 5-foot intervals. Use Simpson's rule to estimate the area of the lot.

11.

x:	0	5	10	15	20	25	30	35	40
y:	60	60	56	52	48	48	52	56	60

12.	x:	0	5	10	15	20	25	30	35	40
	y:	42	48	52	52	54	56	56	60	62

In exercises 13-16, estimate the length of the curve.

13. $y = -\frac{1}{30}x(x - 50)$ on $[0, 50]$ (a 50-yard football punt).

14. $y = 10 + \cosh(x/30)$ on $[-20, 20]$ (the length of a telephone wire).

15. $y = \sin x$ on $[-\pi/6, \pi/6]$ (compare to a straight line).

16. $y = \sin x$ on $[0, 2\pi]$.

In exercises 17-20, find the volume of the solid described.

17. The region in exercise 1 rotated about the x-axis.

18. The region in exercise 2 rotated about the x-axis.

19. The region in exercise 3 rotated about $y = -1$.

20. The region in exercise 4 rotated about the y-axis.

In exercises 21-24, use the given speed to find the distance covered.

21. $s(t) = 36t/\sqrt{t^2 + 2}$ on $[0,20]$ (compare to Example 6).

22. $s(t) = 40t/\sqrt{t^2 + 4}$ on $[0,20]$ (compare to exercise 21).

23. $s(t) = -32t$ on $[0,20]$ (a free fall without air resistance).

24. $s(t) = 130 - 130e^{-t/4}$ on $[0,20]$ (fall with air resistance).

In exercises 25-26, we examine the energy lost to friction in the collision between a ball and striking object. During the collision the ball changes shape, first compressing and then expanding. If x represents displacement in inches and f represents force, the area under the curve $y = f(x)$ is proportional to the energy transferred. In each exercise, $f(x)$ gives the force during the compression of the ball and $g(x)$ gives the force during the expansion of the ball. Thus, the area between the curves $y = f(x)$ and $y = g(x)$ divided by the area under the curve $y = f(x)$ gives the percentage of energy lost due to friction. Compute the percentage. (See *The Physics of Baseball* by R. K. Adair for specific examples.)

25. $f(x) = 25,000x^2 + 10,000x$ and $g(x) = 50,000x^2$.

26. $f(x) = 5000x^2 + 3000x$ and $g(x) = 10,000x^2 + 1000x$.

In exercises 27-30, compute the moment of inertia $I = \frac{1}{3}\int_a^b x^2[f(x)-g(x)]\,dx$ where the region is bounded by $y = f(x)$ on top and $y = g(x)$ on bottom. The regions below are all crude models of baseball bats, and the moment of inertia is a measure of how hard the bat is to swing.

27. $f(x) = \dfrac{1}{2} + \dfrac{x}{30}$, $g(x) = -f(x)$, $a = 0, b = 30$.

28. $f(x) = \dfrac{3}{5} + \dfrac{x}{30}$, $g(x) = -f(x)$, $a = -3, b = 27$.

29. $f(x) = \dfrac{1}{2} + \dfrac{x}{30}$, $g(x) = -f(x)$, $a = 0, b = 32$.

30. Same as exercise 27, but remove the rectangle from $y = -1/4$ to $y = 1/4$ and from $x = 27$ to $x = 30$.

31. If a lawn sprinkler sweeps back and forth at a constant rate, does it provide even coverage? Assume the angle of the sprinkler from the horizontal varies from $\pi/4$ to $3\pi/4$ at the constant rate $d\theta/dt = .2$ rad/sec. Also assume that the water flies $25\sin(2\theta)$ ft horizontally when the sprinkler is at the angle θ. Show that $\theta(t) = .2t + \pi/4$ and $\dfrac{dx}{dt} = 10\cos(.4t + \pi/2)$. Then compute and interpret $\displaystyle\int_0^1 \dfrac{dx}{dt}\,dt$, $\displaystyle\int_1^2 \dfrac{dx}{dt}\,dt$ and $\displaystyle\int_2^3 \dfrac{dx}{dt}\,dt$.

EXPLORATORY EXERCISE

Introduction

If we have points $A = (x_1, y_1)$ and $B = (x_2, y_2)$, the lines from A to the origin and B to the origin are perpendicular if $x_1 x_2 + y_1 y_2 = 0$, since slopes of perpendicular lines (in this case y_1/x_1 and y_2/x_2) multiply out to -1. The most familiar example is $(1,0)$ and $(0,1)$. These points lie on the x- and y-axes, respectively, which we use to represent all other points.

Problems

This exercise extends the idea of perpendicular lines to orthogonal functions. Since the graph of a function includes an infinite number of points, the generalization of $x_1 x_2 + y_1 y_2 = 0$ is an infinite sum equal to 0. Representing the infinite sum as an integral, we define functions f and g to be orthogonal on the interval [a,b] if

$$\int_a^b f(x)g(x)\,dx = 0.$$

Show that the following pairs of functions are orthogonal on $[-1, 1]$: (a) $\sin(\pi x)$ and $\cos(\pi x)$ (b) $\sin(\pi x)$ and $\sin(2\pi x)$ (c) $\cos(\pi x)$ and $\cos(2\pi x)$ (d) 1 and $\sin(\pi x)$.

In fact, any two functions chosen from among $\sin(\pi x)$, $\sin(2\pi x)$, $\sin(3\pi x)$, ... , 1, $\cos(\pi x)$, $\cos(2\pi x)$, ... are orthogonal on $[-1, 1]$. We say that these functions form an orthogonal set of functions on $[-1, 1]$. Show that the set $\{\, 1,\ x,\ x^2,\ x^3,\ ... \,\}$ is not an orthogonal set of functions on $[-1, 1]$.

Find constants c_1-c_7 (not all of which are zero) to make the following an orthogonal set of polynomial functions on $[-1, 1]$:

$$\{\, 1,\ x,\ c_1 x^2 + c_2 x + c_3,\ c_4 x^3 + c_5 x^2 + c_6 x + c_7 g \}$$

HINT: If $f(x) = c_1 x^2 + c_2 x + c_3$ and $g(x) = c_4 x^3 + c_5 x^2 + c_6 x + c_7$, it is necessary that $\int_{-1}^{1} f(x)\,dx$, $\int_{-1}^{1} x f(x)\,dx$, $\int_{-1}^{1} g(x)\,dx$, $\int_{-1}^{1} x g(x)\,dx$ and $\int_{-1}^{1} x^2 g(x)\,dx$ all equal 0. Compute these integrals exactly and find values of the constants (there will be more than one choice) to make all integrals equal zero.

Further Study

The purpose of orthogonal functions is the same as the perpendicular x- and y-axes: to make it easy to represent various functions in a common language. This turns out to be a powerful result in many applications (in case you were wondering what this was doing in an applications section). The Fourier series, discussed in section 6.3, is based on the orthogonal set of sines and cosines above. A general reference is *Orthogonal Transforms for Digital Signal Processing* by Ahmed and Rao.

5.4 Alternative Coordinate Systems

We have emphasized several times that good problem solvers need a variety of skills. For instance, although Simpson's rule and the built-in integration routine on the TI-85 are very powerful, we have seen integrals for which a sequence of Riemann sums provides a better approximation (e.g., Example 5 in section 5.1). Similarly, our standard choice of writing y as a function of x is not appropriate for all integrals of interest. In this section, we will look at two alternatives to using the equation $y = f(x)$ to describe a curve. In some cases, merely treating x as a function of y simplifies calculations. A more fundamental change in perspective is provided by switching to the polar coordinate system.

CHANGE OF INDEPENDENT VARIABLE

In our development of the Riemann sum, we approximated areas by using rectangles of constant width but variable height. When using this scheme, we must keep careful track of the top and bottom of the region we are measuring to determine the heights of the rectangles. Finding the areas of regions without a well-defined top or bottom, such as in Examples 1 and 2, can be awkward.

Example 1. Two Integrals for One Area

Find the area of the region bounded by the graphs of $y = x$, $y = 2 - x$ and $y = 0$. The TI-81 graph of this region is shown in Figure 5.13. This is produced by graphing x and $2 - x$ simultaneously, zooming in to improve the view and doing two Shade commands [on the TI-81, enter Shade(0,X,1,0,1) and Shade(0,2−X,1,1,2); on the TI-85, enter Shade(0,x,0,1) and Shade(0,2−x,1,2)].

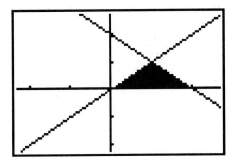

FIGURE 5.13

Geometrically, this problem is simple. The shaded region is a triangle with area $(1/2)(2)(1) = 1$. However, if we wanted to use integrals to represent the area, then we would have to use two integrals to compute the area, since the region is bounded above by $y = x$ for $0 < x < 1$ and by $y = 2 - x$ for $1 < x < 2$. Specifically, we have that

$$\text{Area} = \int_0^1 x \, dx + \int_1^2 (2 - x) \, dx$$

■

Example 2. Two Awkward Integrals for One Area

Sketch the region bounded by the parabolas $x = y^2$ and $x = 2 - y^2$ and represent its area with integrals. Recall from algebra that these parabolas open to the right and left, respectively. Since neither is a function of x, on the calculator you must graph the functions $\sqrt{x}$, $-\sqrt{x}$, $\sqrt{2-x}$ and $-\sqrt{2-x}$ simultaneously. The TI-81 graph (after one zoom in from the standard viewing window) is shown in Figure 5.14.

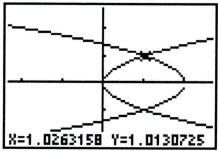

FIGURE 5.14

The region is bounded by the top and bottom of the curve $x = y^2$ for $0 < x < 1$ and by the top and bottom of the curve $x = 2 - y^2$ for $1 < x < 2$, so that

$$\text{Area} = \int_0^1 [\sqrt{x} - (-\sqrt{x})]\, dx + \int_1^2 [\sqrt{2-x} - (-\sqrt{2-x})]\, dx$$

∎

In Examples 1 and 2, we had to use two integrals to find the area of one region. Further, in Example 2 we had to first solve the given equations for y in terms of x to find the functions to be integrated. The thought "there must be a better way" has probably occurred to you, and the form of Example 2 suggests a better way. Why not treat x as a function of y? The general area formula in this case becomes

$$\text{Area} = \int_a^b [g(y) - f(y)]\, dy$$

where $x = g(y)$ defines the right boundary of the region and $x = f(y)$ defines the left boundary of the region. In Figures 5.13 and 5.14, the left and right boundaries of the regions are well-defined, so that this approach should work well in these cases.

Example 3. Areas as Integrals with Respect to y

Compute the areas of Examples 1 and 2 in terms of integration with respect to y. In Example 1, the right boundary of the triangle is the line $y = 2 - x$, which we rewrite as $x = 2 - y$. The left boundary of the triangle is the line $x = y$. The figure extends from $y = 0$ (the lower boundary) to $y = 1$ (the solution of $2 - y = y$). We then have

$$\text{Area} = \int_0^1 [(2 - y) - y] \, dy$$

In Example 2, the right boundary of the region is the curve $x = 2 - y^2$ and the left boundary of the region is the curve $x = y^2$. Since $2 - y^2 = y^2$ if $y = \pm 1$,

$$\text{Area} = \int_{-1}^1 [(2 - y^2) - y^2] \, dy$$

Both integrals are easy to compute by hand (do this now!) and equal 1 and 8/3, respectively.   ∎

POLAR COORDINATES

The circle is one of the most important shapes occurring in nature and in mathematics. It is also one of the few geometrical objects for which we have an exact formula for area. It is ironic, then, that it is very difficult to compute the area of a circle by integrating. First, recall that the circle of radius r centered at the point (a, b) has the equation

$$(x - a)^2 + (y - b)^2 = r^2$$

For the top half of a circle of radius 1 centered at the origin, we have $a = b = 0$ and $r = 1$, so the equation is

$$y = \sqrt{1 - x^2}$$

Then, we get

$$\text{Area} = \int_{-1}^1 \sqrt{1 - x^2} \, dx$$

To compute this exactly, we need the sophisticated (and messy) technique of trigonometric substitution, which you will see in your study of techniques of integration. The news gets worse: it is difficult to even set up the integrals necessary to compute the area of, say, a third of a circle. (Try it!)

The problem is with our approach. As we have defined it, integration is based on sums of areas of rectangles, so that we are trying to fit rectangular pegs into circular holes, so to speak. Of course, you can do this, but it's not pretty. Our first step in improving this situation is to introduce polar coordinates.

The standard rectangular coordinates locate points by measuring a horizontal distance x and vertical distance y from the origin. In polar coordinates, we locate the same point by measuring its distance r from the origin and the angle θ between the line from the origin to the point and the positive x-axis (see Figure 5.15).

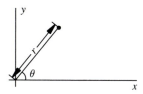

FIGURE 5.15

The angle θ is measured from the positive x-axis as usual: counterclockwise is positive, clockwise is negative, and radian measurement is preferred over degrees. From trigonometry, we get

$$r = \sqrt{x^2 + y^2} \qquad\qquad\qquad x = r \cos \theta$$
$$\tan \theta = y/x \qquad\qquad\qquad\quad y = r \sin \theta$$

The circle $x^2 + y^2 = 4$ then becomes $r^2 = 4$ or $r = 2$ (i.e., the set of all points whose distance from the origin is 2). Also, note that the equation $\theta = \pi/4$ describes the line $y = x$, since $1 = \tan \pi/4 = \tan \theta = y/x$.

The following result is needed to use polar coordinates to compute areas. For a region bounded by the graphs of $\theta = a$, $\theta = b$ and $r = f(\theta)$ as in Figure 5.16, we

$$\text{Area} = \int_a^b \frac{1}{2} f^2(\theta) \, d\theta$$

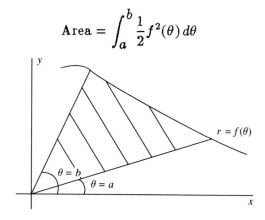

FIGURE 5.16

Example 4. Area of a Circle

Use polar coordinates to compute the area of the circle $r = 1$. In this case, $r = f(\theta) = 1$ and to get the full circle we need θ running from 0 to 2π. Then, we have that

$$\text{Area} = \int_0^{2\pi} (1/2)(1)^2 \, d\theta = \pi$$

∎

Example 5. Area of 1/8-Circle

Compute the area of the region bounded by the graphs of $y = 0$, $y = x$ and the upper portion of the circle $x^2 + y^2 = 4$. This is simply one-eighth of a circle, but the integration in rectangular coordinates is quite ugly. (Try to set it up.) In polar coordinates, the region is bounded by $\theta = 0$, $\theta = \pi/4$ and $r = 2$. We then have

$$\text{Area} = \int_0^{\pi/4} (1/2)(2)^2 \, d\theta = \pi/2$$

∎

Example 6. Area Bounded by a Circle and a Line

Compute the area of the portion of the circle $x^2 + y^2 = 9$ above the line $y = 1$ (see Figure 5.17). In polar coordinates, the equation of the circle is $r = 3$.

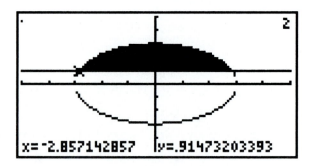

x=-2.857142857   y=.9147320393

FIGURE 5.17

Since $y = r \sin\theta$, we change the equation of the line $y = 1$ into $r \sin\theta = 1$ or $r = 1/\sin\theta$. The intersections occur where $3 = 1/\sin\theta$ or $\sin\theta = 1/3$. Then θ goes from $a = \arcsin(1/3)$ to $b = \pi - \arcsin(1/3)$. Finally, we get

$$\text{Area} = \int_a^b \frac{1}{2}[9 - 1/\sin^2\theta]\,d\theta$$

$$= \int_a^b \frac{1}{2}[9 - \csc^2\theta]\,d\theta$$

$$= \frac{9}{2}(b - a) + \frac{1}{2}[\cot(b) - \cot(a)] = 8.2502026$$

Note that since the TI-81/85 does not have the cotangent function built in, $\cot b$ must be computed as $\cos b/\sin b$. Also, the TI-81/85 does not have the cosecant function built in, so you would need to type $1/\sin^2\theta$ instead of $\csc^2\theta$ to use Simpson's rule or $\boxed{\text{fnInt}}$ to approximate the integral.   ∎

Along with providing a convenient way of computing various areas involving circles, polar coordinates can be used to graph curves that you have probably not yet seen in your studies. The TI-85 provides a built-in utility for drawing graphs in polar coordinates. The TI-81 provides a built-in utility for drawing graphs of parametric equations, which can easily be modified to draw graphs of polar functions. Our modification depends on the relations $x = r\cos\theta$ and $y = r\sin\theta$. Notice that this says that $r = \dfrac{x}{\cos\theta} = \dfrac{y}{\sin\theta}$ and $y = \dfrac{x\sin\theta}{\cos\theta} = x\tan\theta$.

Example 7. Graphing a Spiral

Graph the spiral $r = \theta$. On the TI-81, press $\boxed{\text{MODE}}$, move the cursor down to the Function/Param line, move the cursor over to Param and press $\boxed{\text{ENTER}}$. Next, press $\boxed{\text{Y=}}$. Instead of the function menu with Y_1, Y_2, Y_3, Y_4, you should have lines for X_{1T}, Y_{1T}, X_{2T}, Y_{2T}, X_{3T}, Y_{3T}. Pressing the $\boxed{\text{X|T}}$ key will now produce T's. On the X_{1T} line, enter $T\cos T$. On the X_{2T} line, enter $X_{1T}\tan T$ by retrieving X_{1T} from the Y-vars menu (press $\boxed{\text{Y-vars}}$ 5). Note that we are using the relation $x = r\cos\theta = \theta\cos\theta$ for X_{1T} since we are graphing $r = \theta$. For Y_{1T}, we are using the relation $y = x\tan\theta$ derived above. Note that by doing so, we will not have to adjust Y_{1T} when we change polar functions. We are now ready to adjust the range. Press $\boxed{\text{RANGE}}$ and you will see values for Tmin, Tmax, Tstep, Xmin, Xmax, etc. For this example, we want Tmin=0 and Tmax=9.4248 ($\approx 3\pi$). Finally, press $\boxed{\text{GRAPH}}$ and you should get Figure 5.18.

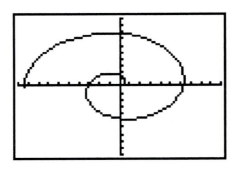

FIGURE 5.18

On the TI-85, press MODE , move the cursor down to the Func/Pol/Param/DifEq line, move the cursor over to highlight "Pol" and press ENTER . Now, press GRAPH $r(\theta) =$ (F1). For $r1$, enter the function θ (use the F1 softkey for θ). Press EXIT RANGE . We want θ Min=0 and θ Max=3π (press 3 π ENTER and the calculator will compute 9.42477796077). Press GRAPH and you will see the spiral of Figure 5.18.   ∎

NOTE: After plotting a polar graph on the TI-81/85, you can graph any number of additional polar graphs without needing to change the mode. The calculator will remain in polar graphing mode until you switch back to function plotting mode. This is done by pressing MODE FUNCTION (or MODE FUNC on the TI-85). The TI-81/85, by the way, does not erase functions when you change modes. After graphing polar graphs, when you return to function mode you will still have whatever functions you last stored in Y1, Y2, ...

Example 8. Area of One Leaf of a Rose

Find the area of one leaf of $r = 6\sin 3\theta$. To graph this on the TI-81, simply change X_{1T} to 6sin3TcosT (do not change Y_{1T}). On the TI-85, set $r1 = 6\sin 3\theta$. You should get the 3-leaf rose seen in Figure 5.19, but traced out more than once (assuming you have the same range values as in Example 7).

To get an idea of the interval of θ's which produces one leaf (these will be our limits of integration), press TRACE . Moving the cursor around the first leaf, it appears that this leaf is traced out with $0 \leq \theta \leq 1.0471976$. In problems involving rectangular coordinates, we would probably mentally round off the upper limit to 1. But in polar coordinates, our first thought should be of multiples of π, and 1.0471976

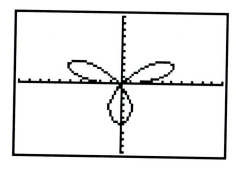

FIGURE 5.19

is about $\pi/3$. We verify this by noting that a leaf starts and stops with $r = 0$ (why is this?), and $r = 0$ if and only if $\theta = 0, \pi/3,...$ Thus,

$$\text{Area} = \int_0^{\pi/3} \frac{1}{2}(2\sin 3\theta)^2 \, d\theta$$

$$= \int_0^{\pi/3} 2\sin^2 3\theta \, d\theta$$

This can be evaluated using the trigonometric identity $2\sin^2 3\theta = 1 - \cos 6\theta$. Thus,

$$\text{Area} = \left[\theta - \frac{1}{6}\sin 6\theta\right]_0^{\pi/3} = \frac{\pi}{3}$$

■

With polar coordinates available, when you encounter a problem (particularly one involving circular geometry) you should now ask yourself whether it is more appropriate to attack the problem using rectangular coordinates (x, y) or using polar coordinates (r, θ).

Exercises 5.4

In exercises 1-6, find the areas of the regions bounded by the given curves using x as a function of y.

1. $y = x$, $y = 2 - x$ and $y = 0$
2. $y = x^2$, $y = 2 - x$, and $y = 0$
3. $x = y^2$ and $x = 8 - y^2$
4. $x = y^2$ and $x = 4$
5. $x^2 + 4y^2 = 4$ and $x = 0$ (right piece)
6. $x = 4 - y^2$ and $x = 0$

In exercises 7-10, find the volume of the solid obtained by rotating the region bounded by the given curves about the x-axis.

7. $y = x$, $y = 2 - x$ and $y = 0$ 8. $x = y^2$ and $x = 2 - y^2$

9. $x = y^2$ and $x = 8 - y^2$ 10. $y = x^2$, $y = 2 - x$ and $y = 0$

In exercises 11-18, sketch a graph of the polar function.

11. $r = \cos 2\theta$ 12. $r = 1 + \sin \theta$

13. $r = 2 \sin \theta$ 14. $r = 2 \cos \theta$

15. $r = 1 + 2 \cos \theta$ 16. $r = 1 + \cos \theta$

17. $r = 2 \cos 3\theta$ 18. $r^2 = 4 \sin \theta$

In exercises 19-24, find the area of the polar region.

19. $r = \cos 2\theta$ (one leaf)   20. $r = 1 + \sin \theta$

21. $r = 2 \sin \theta$ 22. $r = 2 \cos \theta$

23. $r = 1 + 2 \cos \theta$ (inner loop)   24. $r = 1 + \cos \theta$

In exercises 25-30, find the area of the region bounded by the curves.

25. $x^2 + y^2 = 1$, $y = \dfrac{1}{\sqrt{3}}x$ and the positive x-axis

26. $x^2 + y^2 = 1$, $y = -x$ and the positive x-axis

27. $x^2 + y^2 = 9$ above $y = 2$

28. $x^2 + y^2 = 9$ above $y = -2$

29. $(x - 1)^2 + y^2 = 1$ and $x^2 + (y - 1)^2 = 1$

30. $(x - 2)^2 + y^2 = 4$ and $x^2 + (y - 2)^2 = 4$

31. Determine the number of leaves in the roses $r = 6 \sin \theta$, $r = 6 \sin 2\theta$, $r = 6 \sin 3\theta$ and $r = 6 \sin 4\theta$. Conjecture a rule for the number of leaves in $r = 6 \sin n\theta$ for any positive integer n.

32. Determine the number of leaves in the roses $r = 6 \sin \theta/2$, $r = 6 \sin 3\theta/2$, $r = 6 \sin 5\theta/2$ and $r = 6 \sin 7\theta/2$. Conjecture a rule for the number of leaves in $r = 6 \sin n\theta/2$ for any odd integer n.

EXPLORATORY EXERCISE

Introduction

Example 6 in the text and exercises 27-28 refer to the same problem which we call "Fletcher's oil problem." Suppose a cylindrical oil tank (circular cross-sections perpendicular to the ground) has an opening at the top. A measuring stick can be inserted to measure the height of the oil in the tank. If the cylinder has diameter 6 feet, what percentage of the tank is full when the oil has height h?

Problems

We ignore the length of the cylinder (why is this reasonable?) and state the problem as computing $A(h)/9\pi$ where $A(h)$ is the area of that portion of $x^2 + y^2 = 9$ below $y = h$. As in Example 6, if $0 < h < 3$, we compute the area *above* $y = h$ as

$$\int_a^b .5[9 - h^2 \csc^2 \theta]\, d\theta = 4.5(b-a) + h^2[\cot(b) - \cot(a)]$$

where $a = \arcsin(h/3)$ and $b = \pi - \arcsin(h/3)$. The problem is, Fletcher the oil man does not want to calculate arcsines or cotangents. We can do two things for him. First, design a stick with marks at the appropriate heights to indicate $\frac{1}{8}$-tank left, $\frac{1}{4}$-tank left, $\frac{3}{8}$-tank left, etc. Second, come up with a simple rule of thumb to describe how the height relates to the amount of oil left.

Further Study

This problem does not directly relate to other areas of mathematics. However, we hope the reader will devote much further study to the art of making complicated mathematical results useful and easy to understand. Communication between mathematicians and users of mathematics is vitally important.

CHAPTER

6

Sequences and Series

6.1 Sequences

One of the underlying concepts of calculus is that we can often solve complicated problems by generating a sequence of approximations which tend to get closer and closer to the exact solution. We have already seen this idea applied to the study of limits, derivatives and integrals as well as to several rootfinding methods. In this section, we will formally introduce the mathematical notion of a *sequence* of real numbers. Our discussion will serve to unify much of our previous work as well as to lay the foundation for the remainder of the chapter.

Example 1. Conjecturing the Value of a Limit

Conjecture the value of the limit (if it exists) $\lim_{x \to 1} \dfrac{\sqrt{x+1}-2}{x-1}$. We can use FEVAL to generate some function values.

x	1.1	1.01	1.001	1.0001
$f(x)$	.248457	.249844	.249984	.249998

Of course, we also need some function values for $x < 1$.

x	0.9	0.99	0.999	0.9999
$f(x)$	.251582	.250156	.250015	.250001

Based on this evidence, we would feel comfortable conjecturing that the limit exists and equals .25.
∎

Thinking through Example 1, it is natural to focus more on the values of $f(x)$ than on the corresponding x-values. In fact, as long as the x-values approach 1, it does not seem especially important which x's we use. However, we must closely examine the pattern of numbers .248457, .249844, .249984, .249998 and the pattern of numbers .251582, .250156, .250015, .250001. Even though the two sequences of numbers are quite different, they appear to have a common limit of .25.

In the definitions below, we extract the most important features of Example 1 and put them into a general framework. For a more complete discussion of sequences, we refer you to your calculus text.

Roughly speaking, a *sequence* of real numbers is a collection of values, each corresponding to a specific choice of an integer (the *index*). We typically use set notation to describe the sequence a_1, a_2, a_3, ..., writing $\{ a_1, a_2, a_3,...\}$ or more simply $\{a_n\}_{n=1}^{\infty}$.

For example, if $a_n = \dfrac{1}{n^2}$ for $n = 1,2,3,...$, we have the sequence

$$\left\{ \quad 1, \quad \frac{1}{4}, \quad \frac{1}{9}, \quad \frac{1}{16}, \quad \frac{1}{25}, \quad \frac{1}{36}, \quad \frac{1}{49}, \cdots \right\}$$

Note that as we go further and further out in the sequence (i.e., as the index n gets larger and larger), the terms of the sequence are getting closer and closer to 0. In this case, we say that the sequence has the *limit* 0 or that the sequence *converges to 0*.

Loosely speaking, then, a sequence has the limit L if a_n gets closer and closer to L as n gets larger and larger. In this case, we write

$$\lim_{n \to \infty} a_n = L$$

Let's look a little closer to see what this might mean. We should be able to make a_n as close as we like to L , simply by making n large enough. So, given a desired degree of closeness $\epsilon > 0$ (our name for an unspecified positive number) we want to have $|a_n - L| < \epsilon$ for n sufficiently large. More precisely, this means that for every $\epsilon > 0$ there must be an integer N such that if $n > N$, $|a_n - L| < \epsilon$. We take this as our definition of convergence.

Compare this to the definition of the limit of a function given in section 2.3. There, we had said that

$$\lim_{x \to a} f(x) = L$$

if for every $\epsilon > 0$ we could find a $\delta > 0$ so that $|f(x) - L| < \epsilon$ whenever $0 < |x - a| < \delta$. The main difference is the substitution of the condition that $n > N$ for $0 < |x - a| < \delta$. This difference comes from the fact that the sequence a_n is a function defined only for positive integers n and n is tending to infinity. This gives us the flexibility needed to precisely define limits of sequences arising from Newton's method, Riemann sums and many other processes.

Example 2. Finding the Limit of a Sequence

Find the limit of the sequence $\dfrac{1}{2}, \dfrac{2}{3}, \dfrac{3}{4}, \cdots, \dfrac{n}{n + 1}, \cdots$ We first observe that the numbers are all smaller than 1 and appear to be approaching 1, as $n \to \infty$. To prove that $\lim_{n \to \infty} a_n = 1$, we will use the formula $a_n = \dfrac{n}{n + 1}$. From the definition, we need to find an N so that $n > N$ guarantees that

$$\left| \frac{n}{n + 1} - 1 \right| < \epsilon$$

But $\left| \dfrac{n}{n + 1} - 1 \right| = \dfrac{1}{n + 1} < \epsilon$ if $1 < \epsilon(n + 1)$. Solving for n, we get

$$n > \frac{1 - \epsilon}{\epsilon}$$

Thus, if N is any integer greater than $\dfrac{1 - \epsilon}{\epsilon}$, then $|a_n - 1| < \epsilon$ whenever $n > N$, and we have proved that $\lim_{n \to \infty} \dfrac{n}{n + 1} = 1$.

∎

Note the similarity between this proof and that for a limit problem we have already seen, namely $\lim_{x \to \infty} \dfrac{x}{x + 1} = 1$.

Obviously, our proof in Example 2 was dependent on our having a simple formula for a_n. In practice, it is not always possible to simply represent a_n. In that case, we are limited to observing the sequence and conjecturing a value for the limit. All of the examples in this section start with a formula for a_n, so that we may observe several different types of behavior in controlled situations before going on to the more difficult examples to follow.

Example 3. A Limit Involving a Trig Function

Find $\lim_{n \to \infty} \left(1 + \dfrac{\cos \pi n}{n}\right)$, if it exists. We first examine several terms of the sequence and look for a pattern. We find

$$a_1 = 1 + \cos \pi = 0$$

$$a_2 = 1 + \frac{\cos 2\pi}{2} = \frac{3}{2}$$

$$a_3 = 1 + \frac{\cos 3\pi}{3} = \frac{2}{3}$$

$$a_4 = \frac{5}{4} \qquad a_5 = \frac{4}{5}$$

and so on. On the basis of these few terms, you might guess that the sequence is approaching 1. Computing several more terms might serve to convince you of this. We would also be very comfortable conjecturing that

$$\lim_{x \to \infty} \left(1 + \frac{\cos \pi x}{x}\right) = 1 + 0 = 1$$

We use the fact that $|\cos \pi n| = 1$ for every integer n to prove that 1 is the limit. Note that

$$|a_n - 1| = \left|\frac{\cos \pi n}{n}\right| = \frac{1}{n} < \epsilon$$

if $n > 1/\epsilon$. Thus, if N is any integer greater than $1/\epsilon$, then $|a_n - 1| < \epsilon$ whenever $n > N$. ∎

You may have wondered why we seemed to make a distinction between the limit problems $\lim_{x \to \infty} \left(1 + \dfrac{\cos \pi x}{x}\right)$ and $\lim_{n \to \infty} \left(1 + \dfrac{\cos \pi n}{n}\right)$. There is a subtle difference, since in the former x may be any real number while in the latter n may only be an integer. Example 4 illustrates this distinction.

Example 4. A Tricky Limit

Find $\lim_{n \to \infty} (1 + \cos 2\pi n)$, if it exists. First, look at $\lim_{x \to \infty} (1 + \cos 2\pi x)$. We conclude that the limit does not exist, since $\cos 2\pi x$ oscillates between -1 and 1 and, hence, does not approach any limiting value. However, if we list the first few terms of the sequence, we quickly get a different answer, since

$$a_1 = 1 + \cos 2\pi = 2$$
$$a_2 = 1 + \cos 4\pi = 2$$
$$a_3 = 1 + \cos 6\pi = 2$$

and so on. Clearly, $\lim_{n \to \infty} (1 + \cos 2\pi n) = 2$ since $a_n = 2$ for all n. A formal proof is hardly necessary. ∎

In Examples 5 and 6, we examine a pair of sequences which approach 0 as $n \to \infty$. These examples illustrate two important general rules which we will use in section 6.2. The proofs of these results are left for the exercises.

Example 5. Comparing Powers of n

Find $\lim_{n \to \infty} \dfrac{n^2}{n^{5/2} + 3}$ if it exists. We look at several terms of the sequence (use the function evaluation program FEVAL to generate these values):

$$a_1 = .25$$
$$a_{10} = .3132559591$$
$$a_{100} = .0999970001$$
$$a_{1000} = .0316227736$$
$$a_{10000} = .01$$
$$a_{100000} = .0031622777$$
$$a_{1000000} = .001$$

At this point, you might recognize a pattern and conjecture that the sequence is slowly approaching 0. ∎

One of the difficulties in observing limits on the calculator is identifying when a sequence has stopped changing. In this example, $a_{98} \approx .10101$ and $a_{99} \approx .10050$, so it might be tempting to conjecture a limit of .1. It always helps to look at a few more terms of the sequence! It also helps to know some general rules to simplify the thought process. This example illustrates a relatively simple rule. That is, if the terms of $p(n)$ and $q(n)$ are all positive powers of n and the degree (i.e., the largest exponent) of q is larger than the degree of p, then $\lim\limits_{n \to \infty} \dfrac{p(n)}{q(n)} = 0$.

For example, $\lim\limits_{n \to \infty} \dfrac{n^2 + 4}{n^3 + 3n^2 - 1} = 0$ since the degree of $p(n) = n^2 + 4$ is 2 and the degree of $q(n) = n^3 + 3n^2 - 1$ is 3. To verify that this is true, we note that

$$\frac{n^2 + 4}{n^3 + 3n^2 - 1} = \frac{1/n + 4/n^3}{1 + 3/n - 1/n^3}$$

which clearly goes to 0 as $n \to \infty$.

Example 6. Comparing Polynomials and Exponentials

Find $\lim\limits_{n \to \infty} \dfrac{n^2}{e^n}$, if it exists. We compute

$$a_1 = .3678794412$$
$$a_{10} = .004539993$$
$$a_{100} \approx 3.7 \times 10^{-40}$$
$$a_{200} \approx 5.5 \times 10^{-83}$$

It is not difficult to conjecture a limit of 0. Note that both n^2 and e^n tend to infinity as $n \to \infty$. Thus, this limit tells us that e^n must become large faster than n^2 does. In this case, we say that e^n *dominates* n^2, and the general rule is that exponentials with positive exponents will dominate polynomials. ∎

Exercises 6.1

In exercises 1-4, find the limit of the sequence as in Example 2.

1. $\dfrac{3}{1}, \dfrac{5}{2}, \dfrac{7}{3}, \dfrac{9}{4}, \ \dots \ , \dfrac{2n + 1}{n}, \ \dots$

2. $\dfrac{2}{1}, \dfrac{5}{2}, \dfrac{8}{3}, \dfrac{11}{4}, \ \dots \ , \dfrac{3n - 1}{n}, \ \dots$

3. $1, -\dfrac{1}{2}, \dfrac{1}{3}, -\dfrac{1}{4}, \ \dots \ , (-1)^n \dfrac{1}{n}, \ \dots$

4. $2, \dfrac{1}{2}, \dfrac{4}{3}, \dfrac{3}{4}, \ \dots \ , 1 - \dfrac{(-1)^n}{n}, \ \dots$

In exercises 5-10, find the limit of the sequence, if it exists. If the limit exists, find N in terms of ϵ.

5. $a_n = 1 + \dfrac{\sin(\pi n/2)}{n}$

6. $a_n = 1 + \dfrac{\cos(n)}{n}$

7. $a_n = 3 + \sin(\pi n)$

8. $a_n = 2 + \cos(\pi n/2)$

9. $a_n = 2 + \dfrac{(-1)^n}{n}$

10. $a_n = 3 + (-1)^n$

In exercises 11-18, inspect the sequence and conjecture a limit as $n \to \infty$.

11. $a_n = \dfrac{n^3}{n^4 + 5}$

12. $a_n = \dfrac{2n^2 + 4n}{n^3 - 6}$

13. $a_n = \dfrac{3n^3}{8n^2 - 4}$

14. $a_n = \dfrac{5\sqrt{n}}{n + 2}$

15. $a_n = \dfrac{n^2}{e^n}$

16. $a_n = n^3 e^{-n/2}$

17. $a_n = \dfrac{e^n}{n^5 + 4n + 2}$

18. $a_n = \dfrac{e^{-n/2}}{n^2 + 4n + 1}$

In exercises 19-22, use FEVAL (on the TI-81, you can get ! by pressing $\boxed{\text{MATH}}$ 5; on the TI-85, press $\boxed{\text{MATH}}$ $\boxed{\text{PROB}}$ $\boxed{!}$) to conjecture the limit of the sequence, if it exists.

19. $a_n = \dfrac{n^2}{n!}$

20. $a_n = \dfrac{e^n}{n!}$

21. $a_n = \dfrac{n!}{e^{3n}}$

22. $a_n = \dfrac{n^2 e^n}{n!}$

23. Based on your answers in exercises 19-22, which term is dominant, polynomials, exponentials or factorials?

24. Determine the limit of the sequence $\sin(1/x_n)$ for $x_n = 1/n\pi$ and for $x_n = 2/(4n + 1)\pi$. What does this tell you about $\lim\limits_{x \to 0} \sin(1/x)$?

25. Prove that $\lim\limits_{n \to \infty} \dfrac{n^2}{n^{5/2} + 3} = 0$. HINT: $\dfrac{n^2}{n^{5/2} + 3} < n^{-1/2}$ for all n.

26. Prove that $\lim\limits_{n \to \infty} \dfrac{n^2}{e^n} = 0$. HINT: $\dfrac{n^2}{e^n} < \epsilon$ if $n - 2\ln(n) > \ln(1/\epsilon)$.

27. Use the calculator to estimate the limits of $(1 + 1/n)^n$, $(1 + 2/n)^n$ and $(1 + 3/n)^n$ as $n \to \infty$. Compare to e, e^2 and e^3.

EXPLORATORY EXERCISE

Introduction

In applied mathematics, calculations are not typically restricted to real numbers. As you move beyond graphs (where complex numbers have not played a role for us) you will see more and more complex variables. A complex number z may be written as $z = a + bi$ where a and b are real numbers and $i = \sqrt{-1}$. For instance, using the quadratic formula we get the solutions of $x^2 - 2x + 5 = 0$ to be $\dfrac{2 \pm \sqrt{4 - 20}}{2} = 1 + 2i$ and $1 - 2i$. Multiplication of complex numbers depends on the identity $i^2 = -1$. Thus $(1 + 2i)^2 = 1 + 4i + 4i^2 = 1 + 4i - 4 = -3 + 4i$. A useful image is to associate the complex number $a + bi$ with the point (a, b). The TI-85, in fact, does not distinguish between two-dimensional points and complex numbers. Also, the TI-85 uses the same syntax for complex arithmetic as for real arithmetic. In this exercise we will look at the sequence defined by $z_{n+1} = z_n^2 - c$, $z_0 = 0$, where $c = a + bi$ is a complex constant. On the TI-85, enter the following program MAND:

$$:M \wedge 2 - C \boxed{\rightarrow} M \qquad :M$$

On the TI-81, store the following program:

$$:\boxed{\text{Lbl}}\ 1 \quad :X \wedge 2 - Y \wedge 2 - C \boxed{\rightarrow} \boxed{\text{[A]}}\ (1,1) \quad :2X\ Y - D \boxed{\rightarrow} \boxed{\text{[A]}}\ (1,2)$$

$$:\boxed{\text{Disp}}\ \boxed{\text{[A]}} \quad \boxed{\text{[A]}}\ (1,1) \boxed{\rightarrow} X \quad :\boxed{\text{[A]}}\ (1,2) \boxed{\rightarrow} Y \quad :\boxed{\text{Pause}} \quad :\boxed{\text{Goto}}\ 1$$

Problems

Determine the behavior of the sequence for $c = .5, 1, 1.2, 1.5, 3, .2 + .2i, 1 + .2i, 1 + i$ and other values. In general, which c's produce which behavior? To test $c = .5$ on the TI-85, store .5 in C and 0 in M and execute MAND. Press $\boxed{\text{ENTER}}$ to execute the program a second time. Continue pressing $\boxed{\text{ENTER}}$. On the TI-81, store .5 for C and 0 for D (on the TI-81, we start with the complex number $C + Di$), store 0 for X and Y, press $\boxed{\text{MODE}}$ and select FLOAT 5. Then, execute the program MAND. Press $\boxed{\text{ENTER}}$ to continue the program and press $\boxed{\text{ON}}$ when you are done. You should see the sequence converge to about $-.3660$. To test $c = .2 + .2i$ on the TI-85, store $(.2, .2)$ in C and 0 in M, and execute MAND several times. On the TI-81, store .2 in both C and D, 0 in both X and Y and then execute MAND through several iterations. The sequence converges to approximately $-.1864 - .1456i$. Before you jump to an incorrect conclusion, try $c = 1 + i$: the sequence blows up! You should also find values of c for which the sequence eventually alternates between 2 values, values of

c for which the sequence alternates between 4 values, ... The various behaviors are summarized in the remarkable picture below, which shows some of the detail of what is known as the *Mandelbrot set*. The set is displayed using the following rule: if for $c = a + bi$ the sequence blows up, the point (a, b) is colored. The set is sometimes called the "snowman" because it looks like smaller and smaller balls stacked on top of each other (in our picture, the snowman has fallen down). It turns out that points within the same "ball" have the same behavior (for instance, converging to 1 number, or converging to 2 numbers ...). The set is infinitely complicated in the sense that if you zoom in on what appears to be the edge of the set, you reveal more detail and will find what appear to be miniature copies of the set itself!

Further Study

The Mandelbrot set is an example of a *fractal*, of which much has been written recently (see, for example, *The Science of Fractal Images* , ed. by Peitgen and Saupe, Springer-Verlag). The study of sequences such as $z_{n+1} = z_n^2 - c$ belongs to *dynamical systems theory* (see, e.g., *An Introduction to Chaotic Dynamical Systems* by R. Devaney).

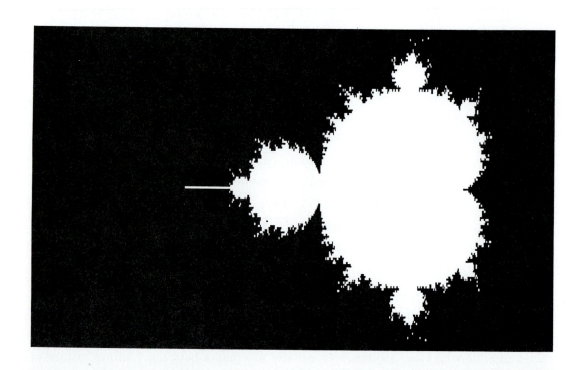

6.2 Infinite Series

Among the many specific sequences which we have already observed in our exploration of calculus are Newton's method approximations and Riemann sums. Both are examples of a special type of sequence, called a *sequence of partial sums*, which we will examine more carefully in this section.

Recall the Newton's method formula for solving the equation $f(x) = 0$:

$$x_{n+1} = x_n - \frac{f(x_n)}{f'(x_n)} \qquad n = 0, 1, 2, \ldots$$

One way to think of this is as

$$x_{n+1} = x_n + c_{n+1} \qquad n = 0, 1, 2, \ldots$$

Here, the new approximation x_{n+1} is the sum of the previous approximation, x_n, plus a so-called *correction term* c_{n+1}, where

$$c_{n+1} = -\frac{f(x_n)}{f'(x_n)}$$

Thus, we have that

$$x_1 = x_0 + c_1$$
$$x_2 = x_1 + c_2 = x_0 + c_1 + c_2$$

and in general,

$$x_n = x_0 + c_1 + c_2 + c_3 + \ldots + c_n$$

We write this in *summation notation* as

$$x_n = x_0 + \sum_{i=1}^{n} c_i$$

If Newton's method succeeds in finding a root, x, then the sequence $\{x_n\}_{n=1}^{\infty}$ converges and

$$x = \lim_{n \to \infty} x_n = x_0 + \lim_{n \to \infty} \sum_{i=1}^{n} c_i$$

which we write as

$$x = x_0 + \sum_{n=1}^{\infty} c_n$$

The last expression in this equation is called an *infinite series* .

We have the following definitions.

Definition For the sequence $\{a_n\}_{n=1}^{\infty}$, the Mth *partial sum* is

$$S_M = \sum_{n=1}^{M} a_n$$

The infinite series $\sum_{n=1}^{\infty} a_n$ is said to *converge* if the sequence of partial sums $\{S_n\}_{n=1}^{\infty}$ converges, in which case we write

$$\sum_{n=1}^{\infty} a_n = \lim_{n \to \infty} S_n$$

If the sequence $\{S_n\}_{n=1}^{\infty}$ does not converge (*diverges*), we say that the infinite series $\sum_{n=1}^{\infty} a_n$ *diverges*.

Example 1. An Infinite Series from Newton's Method

Use Newton's method with an initial guess of $x_0 = 0$ to estimate a root of $x - e^{-x} = 0$. Using the program NEWT from Chapter 4, we get (to 6 decimal places)

$$x_1 = .5$$
$$x_2 = .5663110032$$
$$x_3 = .567143165$$
$$x_4 = .5671432904$$
$$x_5 = .5671432904$$

It appears that Newton's method has converged nicely to a root. We are led to believe this because the iterations get closer and closer together until eventually the displayed digits do not change at all. Said a different way, the correction terms appear to tend to 0. We then expect that the sequence $\{x_n\}_{n=1}^{\infty}$ converges and, hence, that the infinite series $\sum_{n=1}^{\infty} c_n$ also converges, all because the sequence $\{c_n\}_{n=1}^{\infty}$ appears to converge to 0.   ∎

It should be no surprise that we are going to use Example 1 to make what might seem to be a believable conjecture. That is, if $\lim_{n \to \infty} a_n = 0$, we might expect that the infinite series $\sum_{n=1}^{\infty} a_n$ will converge. We should emphasize that this is at least a reasonable expectation, and in fact is often used in casual investigations of series. However, as we will see in the next example, having $\lim_{n \to \infty} a_n = 0$ does *not* guarantee that the infinite series $\sum_{n=1}^{\infty} a_n$ converges. This is one reason that we have so cautiously talked of conjectures throughout the book, while emphasizing the need for double-checking answers. We do not need to look too far to find a counterexample to the preceding conjecture.

Example 2. The Harmonic Series

Show that $\sum_{n=1}^{\infty} \frac{1}{n}$ diverges. Because this series is so important, it is commonly referred to by its name, "harmonic series" (see section 6.3 for the meaning of this name). Clearly $\lim_{n \to \infty} \frac{1}{n} = 0$ and so by our ill-fated conjecture above we might expect that the series converges. In fact, it does seem to converge on the TI-81/85, since on the calculator $1 + 10^{-12}$ "equals" 1 (try this!). But, a clever argument shows that the series, in fact, diverges. First note that

$$\frac{1}{2} + \frac{1}{3} + \frac{1}{4} > 1$$

Also,

$$\frac{1}{5} + \frac{1}{6} + ... + \frac{1}{16} > 1$$

and

$$\frac{1}{17} + \frac{1}{18} + ... + \frac{1}{64} > 1$$

In fact, no matter how large n is, one can show that

$$\frac{1}{n+1} + \frac{1}{n+2} + ... + \frac{1}{4n} > 1$$

(The first n terms of this sum are all at least $\frac{1}{2n}$ and the remaining $2n$ terms are all at least $\frac{1}{4n}$, so the sum is at least $n\frac{1}{2n} + 2n\frac{1}{4n} = \frac{1}{2} + \frac{1}{2} = 1$.) Thus, the sum keeps getting larger and larger without bound (we can always add one more to the sum). We conclude that the series does not converge. ∎

The implication of Example 2 is that simply watching the iterations of Newton's method to see that they get closer together, as in Example 1, does not guarantee the convergence of the series. We would do well to check the conjectured answer $x = .5671432904$ of Example 1 by plugging it back into the function $f(x) = x - e^{-x}$. Here, we get $f(.5671432904) \approx 1 \times 10^{-13}$, which gives us further evidence that we have, in fact, found a good approximation to a root.

NOTE: Our original conjecture should be revised as follows. If $\lim_{n \to \infty} a_n = 0$, then $\sum_{n=1}^{\infty} a_n$ *may or may not* converge. However, if $\lim_{n \to \infty} a_n \neq 0$, then $\sum_{n=1}^{\infty} a_n$ definitely does not converge (it diverges).

It turns out that determining whether a series converges or diverges is just not that easy. We need all of the tests for convergence found in your calculus book, as well as a mental catalog of significant series (such as the harmonic series) and their convergence or divergence properties. Because of loss of significance errors, the calculator is *not* a primary weapon in attacking infinite series. (In fact, the TI-81/85 will suggest to you that the harmonic series converges!) Below, we offer some ways in which the calculator *does* help us in our investigations.

To compute partial sums on the calculator, we can use the built-in sum and sequence commands on the TI-85 (we will describe these commands in Example 3) or the program SUMSEQ given below for the TI-81/85. Recall that the infinite series is the limit of partial sums, so we will want to compute several different partial sums before conjecturing a value for the series. As with limits in Chapter 2, it is the *sequence* of numbers which allows us to make accurate conjectures. A single calculation can be misleading.

TI-81 users should enter the following program. Note that IS> is obtained by pressing PRGM 4.

:SUMSEQ   :Disp "A=,B="   :Input A   :Input B   :0 → S   :Lbl C

:A → X   S+ Y₁ → S   IS> (A,B)   :Goto C   :Disp S

Program Step	Explanation
:SUMSEQ	Name the program.
:$\boxed{\text{Disp}}$ "A=,B="	Prompt for and input A and B,
:$\boxed{\text{Input}}$ A	the starting and stopping
:$\boxed{\text{Input}}$ B	indices for the sum.
:0 $\boxed{\rightarrow}$ S	Set the sum equal to 0.
:$\boxed{\text{Lbl}}$ C	Start the loop.
:A $\boxed{\rightarrow}$ X	Add F(A) to the sum.
:S+ $\boxed{\text{Y}_1}$ $\boxed{\rightarrow}$ S	
:$\boxed{\text{IS>}}$ (A,B)	Repeat the loop if necessary.
:$\boxed{\text{Goto}}$ C	
:$\boxed{\text{Disp}}$ S	Display the partial sum.

To use this program to calculate $\displaystyle\sum_{n=1}^{64} \frac{1}{n}$, first store the function $1/x$ in Y1. Now, execute the program. You will be prompted for A and B (the starting and stopping indices, respectively). When prompted, press 1 $\boxed{\text{ENTER}}$ and 64 $\boxed{\text{ENTER}}$. The calculator returns $S_{64} = 4.743890904$. Other partial sums can be computed by entering different values for A and B. We will indicate how to use the TI-85 commands $\boxed{\text{sum}}$ and $\boxed{\text{seq}}$ in Example 3 below.

You may have already come across geometric series in your regular calculus text. In short, a geometric series is a series for which each term is a constant multiple of the preceding term. That is, these are of the form

$$\sum_{n=0}^{\infty} a \cdot r^n$$

where r is called the *ratio* and a is the first term of the series. One remarkable property of geometric series is that we always know when they converge (when $|r| < 1$). Further, when they converge, they converge to the value $\dfrac{a}{1-r}$. This is extraordinary in that we only rarely know the exact sum of a series. We can use the TI-81/85 to observe the convergence of geometric series.

Example 3. Geometric Series

Find the sum of the geometric series $\frac{3}{4} + \frac{1}{4} + \frac{1}{12} + \frac{1}{36} + ...$ This is a geometric series since each term of the sum is a constant multiple ($r = 1/3$) of the previous term. We write the series as

$$\frac{3}{4}\left[1 + \frac{1}{3} + \left(\frac{1}{3}\right)^2 + ...\right] = \frac{3}{4}\sum_{n=0}^{\infty}\left(\frac{1}{3}\right)^n$$

From the above formula, the series converges to $\frac{3}{4}\left(\frac{1}{1 - 1/3}\right) = \frac{9}{8}$. We can observe the series converge on the TI-81/85. Simply store .75/3^X in Y1 and execute SUMSEQ. At the prompt, enter A=0 and B=100. The partial sum is $S_{100} = 1.125$. Press ENTER to execute the program again, this time entering A=0 and B=200. Again, the partial sum is 1.125. We would suspect that the series has converged, since $S_{100} = S_{200}$. In fact, 1.125 is the exact sum of 9/8. The calculator has given us strong evidence of convergence, but the *proof* is in the derivation of the geometric series formula.

On the TI-85, press MATH MISC and note that *sum* corresponds to F1 and *seq* to F3. Now, press

:sum seq .75/3^N,N,0,100,1)   ENTER

Notice that the first argument (the arguments of sum seq are the expressions set off by commas) is the formula for the N*th* term of the sequence, the second argument is the index of the sequence, the third and fourth arguments are the starting and stopping values for the sum and the last argument is the increment for the sum (if we had used a 2 here, we would have gotten the 0*th*, 2*nd*, 4*th*, ..., 100*th* terms of the sequence added together). After a few seconds, you should see 1.125 displayed. Press ENTRY and change 100 to 200. Pressing ENTER will again produce a partial sum of 1.125. As noted above, this is the exact value of the series.

Of course, not all series will converge as quickly as this one did (i.e., we won't always get so close to an exact value in as few terms of the series as we did here), but for a convergent series, the TI-81/85 helps us to conjecture its value. ∎

Example 4. p-Series

Consider the series $\sum\limits_{n=1}^{\infty} \dfrac{1}{n^2}$. This is an example of another special type of series

called a *p-series* . These are of the form $\sum\limits_{n=1}^{\infty} \dfrac{1}{n^p}$. The present example is the case

where $p = 2$. In your calculus text, you will find that p-series are convergent if $p > 1$ and divergent if $p \leq 1$. Although this helps us easily identify p-series as convergent or divergent (ours is convergent because $p = 2 > 1$), there are no easy ways to compute their sums. On the TI-81/85, enter 1/X∧2 for Y1 and execute program SUMSEQ. At the prompt, enter 1 and 100. You should get $S_{100} = 1.6349839$. Press ENTER to execute SUMSEQ again, and enter 1 and 200. Continue in this way to generate the partial sums listed below.

On the TI-85, press : sum seq 1/N∧2 , N , 1 , 100, 1) ENTER to get S_{100}. Press ENTRY and change 100 to 200. Continuing in this fashion, verify the partial sums:

$$S_{100} = 1.6349839$$
$$S_{200} = 1.639946546$$
$$S_{300} = 1.641606283$$
$$S_{400} = 1.642437189$$
$$S_{500} = 1.642936066$$

This is a slowly converging series, but it appears that the sum is about 1.64. (Amazingly enough, it has been shown that the sum is exactly $\pi^2/6 \approx 1.644934$.) ∎

One of the more useful techniques for determining whether a series converges or diverges is to somehow compare that series with a given series whose convergence or divergence is already known. The two main tools are the following tests.

Comparison Test Suppose that $0 < a_n < b_n$ for all n and that $\sum\limits_{n=0}^{\infty} b_n$ converges.

Then the series $\sum\limits_{n=0}^{\infty} a_n$ also converges. Likewise, if $0 < b_n < a_n$ for all n and $\sum\limits_{n=0}^{\infty} b_n$

diverges, then $\sum\limits_{n=0}^{\infty} a_n$ diverges also.

Limit Comparison Test Suppose that $0 < a_n$, $0 < b_n$, for all n and that $\lim\limits_{n\to\infty} \dfrac{a_n}{b_n} = L$. Then, if $L > 0$, the series $\sum\limits_{n=0}^{\infty} a_n$ and $\sum\limits_{n=0}^{\infty} b_n$ either *both* converge or *both* diverge.

Once you have some experience with these two tests, you will realize that the hardest part of implementing them is finding the right series with which to make a comparison. For the comparison test, it is often hard to see which way the inequality goes for a given prospective comparison series. Your TI-81/85 can be of some help in seeing how to use these tests.

Example 5. The Comparison Test

Determine whether or not the series $\sum\limits_{n=1}^{\infty} \dfrac{\sqrt{n}}{3+n}$ converges. The key here is to notice that for large n, $\dfrac{\sqrt{n}}{3+n} \approx \dfrac{\sqrt{n}}{n} = \dfrac{1}{\sqrt{n}}$. Further, the p-series $\sum\limits_{n=1}^{\infty} \dfrac{1}{\sqrt{n}}$ diverges since $p = 1/2 < 1$. We then conjecture that our series also diverges. However, $\dfrac{\sqrt{n}}{3+n} < \dfrac{1}{\sqrt{n}}$ for all n. Unfortunately, to use the Comparison Test, the inequality needs to go the other way! Don't give up; just compare to a different divergent series. For example, try $\sum\limits_{n=1}^{\infty} \dfrac{1}{n}$. But, is $\dfrac{\sqrt{n}}{3+n} > \dfrac{1}{n}$? Your calculator will help.

Evaluate the function $\dfrac{\sqrt{x}}{3+x} - \dfrac{1}{x}$ at $x = 1$, $x = 2$, ..., until you get positive function values. From the computations, you should notice that

$$\frac{\sqrt{n}}{3+n} > \frac{1}{n} \qquad \text{for} \qquad n > 4$$

Of course, you'll still need to prove that this inequality actually holds for all $n > 4$, but the computations lead us to believe that this is true. For $n > 4$, we get

$$\frac{\sqrt{n}}{3+n} > \frac{\sqrt{n}}{n+n} > \frac{2}{2n} = \frac{1}{n}$$

Thus, by the Comparison Test, we conclude that $\sum\limits_{n=1}^{\infty} \dfrac{\sqrt{n}}{3+n}$ diverges. We note that the Limit Comparison Test also works quite well for this series. (Try this!)   ∎

Recall that an *alternating series* is a series where successive terms are alternately positive and negative. Determining when such a series converges is a fairly simple matter. You will find a version of the following in any standard calculus text.

Alternating Series Test For the alternating series, $\sum\limits_{n=1}^{\infty}(-1)^{n}a_{n}$, if

(i) $a_0 > a_1 > a_2 > ... > a_n > a_{n+1} > ... > 0$ and (ii) $\lim\limits_{n \to \infty} a_n = 0$

then the series converges.

Notice that this is very close to our initial conjecture about a convergence test for series. An additional fact of interest to us is that the sum of a convergent alternating series always lies between successive partial sums. (Think about this: the partial sums will alternately be getting larger and smaller).

Example 6. An Alternating Series

Estimate the sum of the series $\sum\limits_{n=1}^{\infty}(-1)^{n}\dfrac{\sqrt{n}}{3+n}$. First notice that the series converges, by the alternating series test. (Make sure that you check the details of this.) Using the SUMSEQ program on the TI-81/85 or the $\boxed{\text{sum}}$ $\boxed{\text{seq}}$ commands on the TI-85, we get the partial sums

$$S_{100} = -.0632138116$$
$$S_{500} = -.0894268795$$
$$S_{1000} = -.0958830357$$
$$S_{2000} = -.1004810111$$
$$S_{2001} = -.122802637$$

Since the sum of this alternating series is between S_{2000} and S_{2001}, we know that the sum is between $-.100$ and $-.123$. This leaves a wide range of possible values, especially considering that we have summed the first 2001 terms. Without adding more terms, we may conclude only that the limit of this slowly converging series is *about* $-.11$. ∎

We conclude this section with two powerful tests of convergence. First, recall that the series $\displaystyle\sum_{n=1}^{\infty} a_n$ is said to *converge absolutely* if the series of absolute values,

$\displaystyle\sum_{n=1}^{\infty} |a_n|$, converges.

The Root Test Suppose that $\displaystyle\lim_{n \to \infty} \sqrt[n]{|a_n|} = L$. Then, if $L < 1$, the series converges absolutely. If $L > 1$, the series diverges. Finally, if $L = 1$, the test yields no information.

Example 7. The Root Test

Determine whether or not $\displaystyle\sum_{n=1}^{\infty} \frac{n^3}{3^n}$ converges. Note that because of the presence of the term 3^n, it is reasonable to try the Root Test. We need to compare $\displaystyle\lim_{n \to \infty} \sqrt[n]{n^3/3^n} = \lim_{n \to \infty} n^{3/n}/3$ to 1. But, is $\displaystyle\lim_{n \to \infty} n^{3/n} < 3$? We can use our function evaluation program FEVAL to explore the behavior of the sequence $n^{3/n}$. Store the function XΛ(3/X) as Y1 and generate the following values:

$$
\begin{array}{lll}
1.1481536215 & \text{for} & \text{X}=100 \\
1.02093948371 & \text{for} & \text{X}=1000 \\
1.002766923 & \text{for} & \text{X}=10000
\end{array}
$$

and so on. It would seem reasonable to conjecture that $\displaystyle\lim_{n \to \infty} n^{3/n} < 3$, in which case the series converges by the Root Test. In fact, the limit in question is 1. (You can show this by looking at the natural logarithm of the expression.) Now that we know that the series converges, we can use SUMSEQ on the TI-81/85 or $\boxed{\text{sum}}$ $\boxed{\text{seq}}$ on the TI-85 to compute some partial sums in an effort to approximate the series. In this case, we get $S_{100} = S_{200} = 4.125$.   ∎

The Ratio Test may be the one test most frequently used in applications.

The Ratio Test Suppose that $\displaystyle\lim_{n \to \infty} \left| \frac{a_{n+1}}{a_n} \right| = L$. If $L < 1$, then the series $\displaystyle\sum_{n=1}^{\infty} a_n$ converges absolutely. If $L > 1$, the series diverges. Finally, if $L = 1$, the test yields no information.

Example 8. The Ratio Test

Determine whether or not $\sum_{n=0}^{\infty} \frac{1}{n!}$ converges. We first note that series involving factorials can often be examined using the ratio test. Here, we have

$$\lim_{n \to \infty} \left| \frac{a_{n+1}}{a_n} \right| = \lim_{n \to \infty} \frac{n!}{(n+1)!}$$

$$= \lim_{n \to \infty} \frac{n!}{(n+1)n!}$$

$$= \lim_{n \to \infty} \frac{1}{n+1} = 0 < 1$$

The ratio test then says that the series converges absolutely. The next reasonable question to ask is, "What does the series converge to?" Ordinarily, this remains unknown. We will again use our program SUMSEQ on the TI-81/85 or [sum] [seq] on the TI-85 to answer the question. We get:

$$S_{50} = 2.71828182846$$

$$S_{60} = 2.71828182846$$

and so on. In fact, $S_{1000} = 2.71828182846$. Do you recognize this number? You should. It's the irrational number e (at least a 12-digit approximation of e). ∎

Series can be quite tricky to deal with in practice. However, armed with your TI-81/85 and a full array of convergence tests, you can readily discover when they converge and compute approximations to the values to which they converge.

Exercises 6.2

In exercises 1-18, estimate the sum of the infinite series, if it converges.

1. $2 + \frac{1}{2} + \frac{1}{8} + \ldots + \frac{2}{4^n} + \ldots$

2. $1 - \frac{1}{2} + \frac{1}{4} + \ldots + (-1)^n \frac{1}{2^n} + \ldots$

3. $\frac{2}{3} + 1 + \frac{3}{2} + \ldots + \frac{2(3)^n}{3(2)^n} + \ldots$

4. $\frac{3}{4} - \frac{1}{4} + \frac{1}{12} + \ldots + (-1)^n \frac{3}{4(3)^n} + \ldots$

5. $-1 + \frac{2}{3} - \frac{4}{9} + \ldots - (-1)^n \frac{2^n}{3^n} + \ldots$

6. $-2 - \frac{2}{3} - \frac{2}{9} - \ldots - \frac{2}{3^n} - \ldots$

7. $\sum_{n=1}^{\infty} \frac{1}{n^3}$

8. $\sum_{n=1}^{\infty} \frac{1}{n^{2/3}}$

9. $\displaystyle\sum_{n=1}^{\infty}(-1)^n \frac{3}{n}$

10. $\displaystyle\sum_{n=1}^{\infty}(-1)^{n+1} \frac{4n}{n+1}$

11. $\displaystyle\sum_{n=2}^{\infty} \frac{4}{n^2}$

12. $\displaystyle\sum_{n=3}^{\infty} \frac{-2}{n^{1.2}}$

13. $\displaystyle\sum_{n=1}^{\infty} \frac{\cos(\pi n)}{n}$

14. $\displaystyle\sum_{n=1}^{\infty} \frac{\sin(\pi n/2)}{2n}$

15. $\displaystyle\sum_{n=1}^{\infty} \frac{2}{\sqrt{n}}$

16. $\displaystyle\sum_{n=2}^{\infty} \frac{4}{n\sqrt{n}}$

17. $\displaystyle\sum_{n=1}^{\infty}(-1)^n \frac{3n}{\sqrt{n^2+2}}$

18. $\displaystyle\sum_{n=1}^{\infty}(-1)^n \frac{5}{\sqrt{n^2+4}}$

In exercises 19-22, use the Comparison Test to determine whether or not the series converges.

19. $\displaystyle\sum_{n=1}^{\infty} \frac{2}{n^2-3}$

20. $\displaystyle\sum_{n=1}^{\infty} \frac{n+1}{n^3-5}$

21. $\displaystyle\sum_{n=1}^{\infty} \frac{3}{\sqrt{n}+2}$

22. $\displaystyle\sum_{n=1}^{\infty} \frac{n-1}{n^2+3}$

In exercises 23-26, use the Root Test or the Ratio Test to determine whether or not the series converges.

23. $\displaystyle\sum_{n=1}^{\infty} \frac{n^2}{6^n}$

24. $\displaystyle\sum_{n=1}^{\infty} \left(\frac{2}{3+n}\right)^n$

25. $\displaystyle\sum_{n=1}^{\infty} \frac{n2^n}{n^{2n}}$

26. $\displaystyle\sum_{n=1}^{\infty} \frac{n^2 2^n}{e^n}$

27. In exercise 26 of section 5.1, you were asked to conjecture whether or not $\displaystyle\int_0^1 \sin(1/x)\,dx$ exists. To argue that it does exist, we will use infinite series.

Since $\sin(1/x) = 0$ if $x = 1/\pi,\ 1/2\pi,\ 1/3\pi,\ \ldots$ compute $\displaystyle\int_{1/\pi}^1 \sin(1/x)\,dx$,

$\displaystyle\int_{2/\pi}^{1/\pi} \sin(1/x)\,dx,\ \int_{3/\pi}^{2/\pi} \sin(1/x)\,dx,\ \ldots$ Each individual integral exists, but does the sum of the integrals exist? Compute the 3 integrals above to see if a pattern develops. Note that they alternate signs, so the infinite series converges if the sequence $a_n = \int_{1/n\pi}^{1/(n+1)\pi} \sin(1/x)\,dx$ tends to 0. Using $|\sin(1/x)| \leq 1$, show that $|a_n|$ tends to 0.

28. What are the odds of winning a deuce game in tennis? In this situation, a player wins the game by winning two points in a row. If each player wins one point, the deuce starts over. If player A wins 60% of the points, A wins both points with probability .36 and the points are split with probability .48. Of the 48% split, A wins both of the next two points with probability .36 and they split points again with probability .48. Argue that player A wins the game with probability $.36 + (.48)(.36) + (.48)^2(.36) + (.48)^3(.36) + ...$ and compute the sum.

29. A basketball player makes 90% of his free throws. How many would you expect to be made before the first miss? If the $(n+1)$st is the first miss, there are n made and then 1 missed, which occurs with probability $(.9)^n(.1) = p(n+1)$. The expected number of free throws made is $\sum_{n=0}^{\infty} np(n+1)$. Estimate the sum of this series.

EXPLORATORY EXERCISE

Introduction

Infinite series are useful in many situations. In exercise 27, we saw an infinite series of integrals. We will see in the next section that infinite series of functions are important. We take a quick look at such a series in this exercise.

Problems

Start by doing long division to get $\dfrac{1}{1+x^2} = 1 - x^2 + x^4 - x^6 + ...$. This equation makes sense only if the infinite series converges. Show that the series converges if $-1 < x < 1$ and diverges otherwise. Now integrate the equation term by term to get $\arctan(x) = x - x^3/3 + x^5/5 - x^7/7 + ...$. Again, the equation makes sense only if the series converges. Show that the series converges if $-1 < x \leq 1$ and diverges otherwise. Finally, determine all x's for which the following series converge, and estimate the value of the series for three different x's: $\sum_{n=1}^{\infty} \left(\dfrac{x}{2}\right)^n$ and $\sum_{n=1}^{\infty} \dfrac{x^n}{n!}$.

Further Study

It turns out that the above results are not unusual. *Power series* (series involving powers of x) have what is called a *radius of convergence*. If a power series converges on the interval $(0, r)$ then it also converges on $(-r, 0)$. The convergence at the endpoints $x = -r$ and $x = r$ may differ from problem to problem. Your calculus book has more information on power series.

6.3 Series Representations of Functions

Much of modern technology is based on an understanding of how to break down a complicated task into simpler components. For example, a sound engineer might turn up the bass to produce the desired tone in a recording. Internally, the recording equipment executes a complicated process of amplification and deamplification to suppress unwanted recording noise. A signal processing engineer might electronically boost the high frequencies in a digitized photograph, transforming a blurred image into a sharp picture. In all of these scenarios, a target function (e.g., the desired sound or picture) is obtained by assembling component parts in the proper proportions.

Infinite series are often used to represent some quantity of interest. The partial sums are then used to approximate the sum of the series, providing increasingly accurate approximations as the number of terms summed increases. For instance, five iterations of Newton's method may give us a good approximation of the solution of an equation. The sixth iteration (i.e., adding one more term to the partial sum) will generally give a better approximation. Finally, the exact solution equals the limit of the Newton approximations (i.e., the sum of the series).

In this section, we will extend the notion of series representations from those for single numbers to those for functions. That is, given a function $y = f(x)$, we will approximate it as a sum of simpler functions (e.g., polynomials), with the property that the approximation improves as we add more terms to the sum. This may sound like an ambitious project, but there are numerous important applications based on such series representations of functions. For instance, we shall see how a music synthesizer uses a series of pure tones to imitate a particular musical instrument. We will discuss the two most prominent series representations, Taylor series and Fourier series.

TAYLOR SERIES

We have already seen some examples of *Taylor polynomials*, although we have sometimes called them by other names. For instance, what is the best straight-line approximation of $y = x^2 - 1$? If we are especially interested in maintaining accuracy near the point $(1,0)$ we would choose the tangent line to the curve at $(1,0)$, namely $y = 2x - 2$ (see Figure 6.1a for a TI-85 graph). If we are more interested in accuracy near the point $(2,3)$, we would choose the tangent line at $(2,3)$ given by $y = 4x - 5$ (see Figure 6.1b for a TI-85 graph). These tangent lines are examples of Taylor polynomials of degree 1.

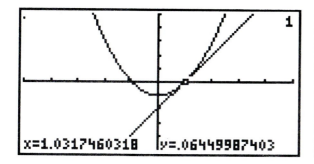

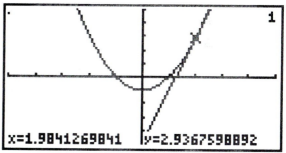

FIGURE 6.1a   FIGURE 6.1b

Of course, few things in life are linear (i.e., follow straight lines) and we need to develop better approximations than those obtained from tangent lines. We must ask more difficult questions. For a function $f(x)$, which quadratic (2nd-order polynomial) function best approximates f near some point? What is the best cubic approximation? What is the best 4th-order approximation? The following definition provides some clues.

Definition The *Taylor polynomial* of degree n approximating the function $f(x)$ and centered about $x = a$ is

$$P_n(x) = c_0 + c_1(x - a) + c_2(x - a)^2 + c_3(x - a)^3 + \ldots + c_n(x - a)^n$$

where the coefficients are given by $c_i = \dfrac{f^{(i)}(a)}{i!}, \quad i = 0, 1, 2, \ldots, n.$

Example 1. Computing Taylor Polynomials

Compute the Taylor polynomials of degrees 1, 2 and 3 centered about $x = 0$ for $f(x) = e^x - 1$. First, note that

$$f'(x) = f''(x) = f'''(x) = \dots = f^{(n)}(x) = e^x$$

We then compute the coefficients $c_0 = f(0) = 0$, $c_1 = f'(0) = 1$, $c_2 = f''(0)/2 = 1/2$ and $c_3 = f'''(0)/3! = 1/6$. Thus,

$$P_1(x) = 0 + 1(x - 0) = x$$

Note that this is the tangent line at $(0,0)$ as seen in Figure 6.2a [the TI-85 display of $P_1(x)$ and $f(x)$]. Similarly,

$$P_2(x) = 0 + 1(x - 0) + \frac{1}{2}(x - 0)^2 = x + \frac{x^2}{2}$$

[see Figure 6.2b for the TI-85 graph of $P_2(x)$ and $f(x)$].

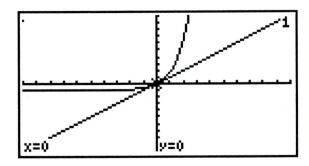   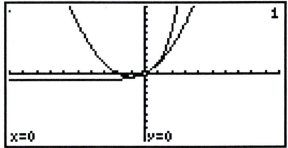

FIGURE 6.2a   FIGURE 6.2b

Finally, $P_3(x) = 0 + 1(x - 0) + \frac{1}{2}(x - 0)^2 + \frac{1}{6}(x - 0)^3 = x + \frac{x^2}{2} + \frac{x^3}{6}$. Figure 6.3 shows the TI-85 graph of $P_3(x)$ and $f(x)$.

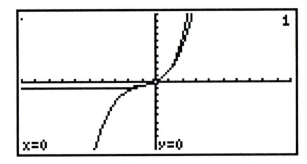

FIGURE 6.3

You should notice that $P_2(x) = P_1(x) + \dfrac{x^2}{2}$ and $P_3(x) = P_2(x) + \dfrac{x^3}{6}$. Taylor polynomials of higher order (i.e., higher degree) build on the Taylor polynomials of lower order. This can save us keystrokes on the TI-81/85 if we enter Y1=X, Y2=Y1+X∧2/2 and Y3=Y2+X∧3/6.

Also, notice in Figures 6.2-6.3 that all three polynomials are close to the graph of $e^x - 1$ near $x = 0$. However, P_2 remains close for a wider domain of x's than P_1, and P_3 in turn remains close longer than P_2. That is, our approximations of $e^x - 1$ are improving as the degree of the approximating polynomial gets larger, just as we wanted! ∎

Example 2. Taylor Polynomials for sin(x)

Repeat Example 1 for $f(x) = \sin(x)$. Note that $f'(x) = \cos(x), f''(x) = -\sin(x)$ and $f'''(x) = -\cos(x)$. We then compute the coefficients $c_0 = \sin 0 = 0$, $c_1 = \cos 0 = 1$, $c_2 = -\sin 0/2 = 0$ and $c_3 = -\cos 0/6 = -1/6$. Then $P_1(x) = x$ is the tangent line seen in Figure 6.4a [the TI-81 graph of $P_1(x)$ and $f(x)$]. We also have $P_2(x) = x$, since $c_2 = 0$. Finally, $P_3(x) = x - \dfrac{x^3}{6}$ [see Figure 6.4b for the TI-81 graph of $P_3(x)$ and $f(x)$]. Notice that here $P_2(x) = P_1(x)$. This points out that the Taylor polynomial P_n is actually a polynomial of degree $at\ most\ n$. You can think about the fact that $P_2 = P_1$ in the following way. You cannot draw a parabola that approximates $y = \sin x$ any better (at least near $x = 0$) than the straight line shown in Figure 6.4a. (Try this for yourself.) ∎

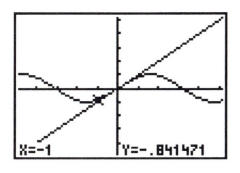

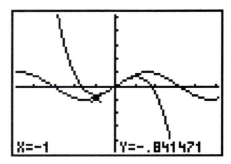

FIGURE 6.4a   FIGURE 6.4b

Taylor polynomials can be used to approximate all sorts of quantities. For example, in the exercises in Chapter 3, we saw how to use them to approximate the solution of a differential equation. Taylor polynomials can also be used to approximate the value of a definite integral.

Example 3. Estimating the Value of an Integral

Use the Taylor polynomial $P_3(x)$ found above to estimate $\int_0^1 \sin(x^2)\,dx$. Since

$P_3(x)$ approximates $\sin x$, we write $\sin x \approx x - \dfrac{x^3}{6}$ and hence

$$\sin(x^2) \approx x^2 - \frac{(x^2)^3}{6} = x^2 - \frac{x^6}{6}$$

Integrating, we get

$$\int_0^1 \sin(x^2)\,dx \approx \int_0^1 \left(x^2 - \frac{x^6}{6}\right) dx = \frac{1}{3} - \frac{1}{42} = .30952$$

The exact value is .31027.   ■

Example 4. Higher-Order Taylor Polynomials

Compute $P_8(x)$ and $P_{16}(x)$ centered about $x = 0$ for $f(x) = \sin x$ and compare their graphs to that of $y = \sin x$. Extending our work in Example 2, we compute $P_4(x) = x - x^3/6 = x - x^3/3!$, $P_5(x) = P_6(x) = x - x^3/3! + x^5/5!$ and $P_7(x) = P_8(x) = x - x^3/3! + x^5/5! - x^7/7!$. Enter P_8 as Y1 (note that the factorial function ! is obtained by pressing $\boxed{\text{MATH}}$ 5 on the TI-81 or $\boxed{\text{MATH}}$ $\boxed{\text{PROB}}$ $\boxed{!}$ on the TI-85). At this point, the pattern of terms should be apparent, and we can enter Y2=Y1+x^9/9! − x^{11}/11! + x^{13}/13! − x^{15}/15! . The TI-85 graphs of $y = \sin x$ and $y = P_8(x)$ are shown in Figure 6.5a, and the graphs of $y = \sin x$ and $y = P_{16}(x)$ are shown in Figure 6.5b.

Note that, to within the resolution of the calculator's graphics display, $P_{16}(x) = \sin x$ for all x between about $x = -6.5$ and $x = 6.5$.   ■

We have so far accomplished at least part of our goal: we now have a means of finding increasingly accurate polynomial approximations to a given function. The final question is: is the function given exactly by the limit of this sequence of approximations? Theorem 6.1 tells us when the answer is yes. First, we give a name to this limit.

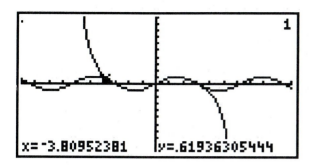
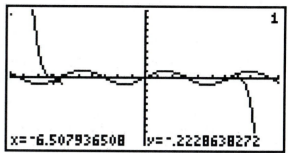

FIGURE 6.5a   FIGURE 6.5b

Definition For a function $f(x)$, the series $\sum_{n=0}^{\infty} \dfrac{f^{(n)}(a)}{n!}(x-a)^n$ is called the *Taylor series expansion* for $f(x)$ about $x = a$.

Theorem 6.1 Suppose that the function f has derivatives of all orders (i.e., f', f'', f''', ... all exist) in some open interval containing $x = a$. If the Taylor series $\sum_{n=0}^{\infty} \dfrac{f^{(n)}(a)}{n!}(x-a)^n$ converges and $\lim_{n \to \infty} \dfrac{f^{(n)}(c)}{n!}(x-a)^n = 0$ for each x and c in the interval, then the Taylor series converges to $f(x)$ for all x in the interval.

Thus, we may compute the Taylor series and then determine for which x's the series converges. If the series converges in an interval containing x and the limit equals 0, the series converges to $f(x)$. We do not have to worry about the series converging to the wrong value.

Example 5. Using Taylor Series to Find Limits

Recall that in Chapter 2, we discussed the problem of loss of significance errors in computing limits. Here, we use Taylor series to argue that

$$\lim_{x \to 0} \frac{1 - \cos x}{x^2} = \frac{1}{2}$$

Recall from Example 5 of section 2.2 that the TI-81/85 cannot accurately compute the values of the above function for values of x very close to 0. As an alternative to simply typing in the function, we first compute the Taylor series $1 - \dfrac{x^2}{2} + \dfrac{x^4}{4!} - \dfrac{x^6}{6!} + \cdots$ for $\cos x$. Identifying $a_n = \dfrac{(-1)^n x^{2n}}{(2n)!}$, we can use the Ratio Test to prove that the series converges for all x. Further, $\lim\limits_{n \to \infty} \dfrac{f^{(n)}(c)}{n!} x^n = 0$ for all x, since $|\cos c| \leq 1$ and $|\sin c| \leq 1$ for all c. It is then meaningful to write

$$\cos x = 1 - \frac{x^2}{2} + \frac{x^4}{4!} - \frac{x^6}{6!} + \ldots$$

It follows that

$$1 - \cos x = \frac{x^2}{2} - \frac{x^4}{4!} + \frac{x^6}{6!} + \ldots$$

and

$$\frac{1 - \cos x}{x^2} = \frac{1}{2} - \frac{x^2}{4!} + \frac{x^4}{6!} + \ldots \quad (x \neq 0)$$

We then conclude that $\lim\limits_{x \to 0} \dfrac{1 - \cos x}{x^2} = \lim\limits_{x \to 0} \left[\dfrac{1}{2} - \dfrac{x^2}{4!} + \dfrac{x^4}{6!} + \ldots \right] = \dfrac{1}{2}.$

∎

The reader should beware that our calculations are "formal." That is, they look right, but we have not taken the care to show that each step in the derivation is legal and meaningful (for instance, for which x's does the series $\dfrac{1}{2} - \dfrac{x^2}{4!} + \dfrac{x^4}{6!} + \ldots$ converge? Can we compute the limit of the series simply by plugging in $x = 0$?).

FOURIER SERIES

Among the many benefits of Taylor series is that they enable us to calculate values such as $\sin(1)$ using only addition and multiplication. This is the language of humans and our computers. However, nature's language is often not arithmetic. Nature often comes to us in sine waves: sight and sound, for instance, are essentially wave phenomena. In applications involving waves, sines and cosines are typically simpler and more natural than polynomials. We are then led to solve the following problem. Given a function $f(x)$, represent f as a series of sines and cosines.

We start with a simple example. Suppose that we know in advance that a function is the sum of a few sines and cosines. For instance, suppose

$$f(x) = a_1 \cos \pi x + a_2 \cos 2\pi x + b_1 \sin \pi x + b_2 \sin 2\pi x$$

Given $f(x)$, how can we determine the constants a_1, a_2, b_1 and b_2? The solution may not be obvious, but we only need integration to understand it. First, multiply the above equation by $\cos \pi x$ and then integrate from -1 to 1. We get

$$\int_{-1}^{1} f(x) \cos \pi x \, dx = a_1 \int_{-1}^{1} \cos^2 \pi x \, dx + a_2 \int_{-1}^{1} \cos 2\pi x \cos \pi x \, dx$$

$$+ b_1 \int_{-1}^{1} \sin \pi x \cos \pi x \, dx + b_2 \int_{-1}^{1} \sin 2\pi x \cos \pi x \, dx$$

This looks like a mess, but evaluate the integrals. You should find that the first integral on the right is 1 and the other 3 integrals are 0! Thus,

$$\int_{-1}^{1} f(x) \cos \pi x \, dx = a_1$$

and we have solved for a_1. How can we find a_2? Multiply the original equation by $\cos 2\pi x$ and integrate from -1 to 1. Here, we get

$$\int_{-1}^{1} f(x) \cos 2\pi x \, dx = a_1 \int_{-1}^{1} \cos \pi x \cos 2\pi x \, dx + a_2 \int_{-1}^{1} \cos^2 2\pi x \, dx$$

$$+ b_1 \int_{-1}^{1} \sin \pi x \cos 2\pi x \, dx + b_2 \int_{-1}^{1} \sin 2\pi x \cos 2\pi x \, dx$$

Again, all but one of the integrals on the right side are 0 and we get

$$\int_{-1}^{1} f(x) \cos 2\pi x \, dx = a_2$$

You should be able to supply the details behind the remaining formulas:

$$\int_{-1}^{1} f(x) \sin \pi x \, dx = b_1$$

$$\int_{-1}^{1} f(x) \sin 2\pi x \, dx = b_2$$

We now present the general result.

For a function f, define the *Fourier series* of f on the interval $[-L, L]$ by

$$\frac{a_0}{2} + \sum_{n=1}^{\infty} \left(a_n \cos \frac{n\pi x}{L} + b_n \sin \frac{n\pi x}{L} \right)$$

By essentially the same process as that illustrated above, we find that

$$a_n = \frac{1}{L} \int_{-L}^{L} f(x) \cos(n\pi x/L)\, dx$$

and

$$b_n = \frac{1}{L} \int_{-L}^{L} f(x) \sin(n\pi x/L)\, dx$$

The constants a_n, $n = 0, 1, 2, \ldots$ and b_n, $n = 1, 2, \ldots$ are called the *Fourier coefficients* of f. For details of this, see Churchill and Brown, *Fourier Series and Boundary Value Problems*.

Theorem 6.2 Suppose that f and f' are continuous on the interval $[-L, L]$ except for possibly a finite number of jump discontinuities. Then, the Fourier series for f on $[-L, L]$ converges to $f(x)$ at all points where f is continuous.

This is indeed a nice result! The function f does not even have to be continuous for the series to converge. However, Fourier series are often very slowly converging (i.e., it takes many terms of the series to obtain a reasonable approximation), as we shall see in the examples to follow.

Example 6. Fourier Series

Compute the Fourier series for $f(x) = x^2$ on $[-1, 1]$ and graph the 4th and 8th partial sums of the series. We compute the coefficients $a_n = \int_{-1}^{1} x^2 \cos n\pi x\, dx$ and $b_n = \int_{-1}^{1} x^2 \sin n\pi x\, dx$ using integration by parts. We get

$$a_n = \left[x^2 \frac{1}{n\pi} \sin n\pi x + 2x \frac{1}{n^2 \pi^2} \cos n\pi x - 2 \frac{1}{n^3 \pi^3} \sin n\pi x \right]_{-1}^{1}$$

$$= \frac{4(-1)^n}{n^2 \pi^2} \qquad \text{if} \qquad n \neq 0$$

$$b_n = \left[-x^2 \frac{1}{n\pi} \cos n\pi x + 2x \frac{1}{n^2 \pi^2} \sin n\pi x + 2 \frac{1}{n^3 \pi^3} \cos n\pi x \right]_{-1}^{1} = 0$$

Also, $a_0 = \displaystyle\int_{-1}^{1} x^2 \, dx = \dfrac{2}{3}$. The Fourier series for f is then

$$\frac{1}{3} + \frac{4}{\pi^2}\left[-\cos \pi x + \frac{1}{4}\cos 2\pi x - \frac{1}{9}\cos 3\pi x + \frac{1}{16}\cos 4\pi x + \dots\right]$$

The TI-85 graphs of $y = x^2$, together with the 4th and 8th partial sums, are shown in Figures 6.6a and 6.6b, respectively (these were obtained by zooming in once from the standard viewing window).

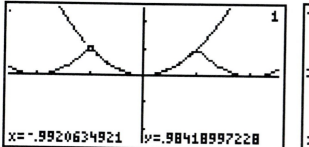

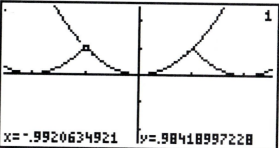

FIGURE 6.6a   FIGURE 6.6b

Note that in both cases, the curves are nearly indistinguishable on the interval $[-1, 1]$. Although the curves are not particularly close outside of $[-1, 1]$, this is not a deficiency, since our only intention in finding the Fourier series expansion was to find an approximation valid on $[-1, 1]$.   ∎

Example 7. Fourier Series

Repeat Example 6 for $f(x) = x$. This time,

$$a_n = \int_{-1}^{1} x \cos n\pi x \, dx = 0 \text{ for } n = 0, 1, 2, \dots$$

$$b_n = \int_{-1}^{1} x \sin n\pi x \, dx$$

$$= \left[-x\frac{1}{n\pi}\cos n\pi x + \frac{1}{n^2\pi^2}\sin n\pi x\right]_{-1}^{1}$$

$$= \frac{-2(-1)^n}{n\pi}$$

The Fourier series is then

$$\frac{2}{\pi} \left[\sin \pi x - \frac{1}{2} \sin 2\pi x + \frac{1}{3} \sin 3\pi x - \frac{1}{4} \sin 4\pi x + ... \right]$$

The TI-85 graphs of $y = x$, together with the 4th and 8th partial sums, are shown in Figures 6.7a and 6.7b, respectively (we have zoomed in once from the standard viewing window).   ∎

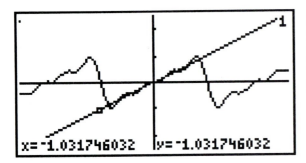

FIGURE 6.7a   FIGURE 6.7b

Figures 6.6a-6.7b may surprise you. It really is possible to approximate parabolas and straight lines with sums of sines and cosines. The computation of the Fourier coefficients was not easy. With practice, the symmetry tricks (e.g., in Example 7, a_n can be seen to be 0 because the integrand, $x \cos n\pi x$, is an odd function and the integration is over a symmetric interval) and integration by parts will become routine.

The series derived in Example 7 is very important in the design of music synthesizers. On an oscilloscope, each musical instrument has an identifying waveform (see Figures 6.8a and 6.8b for the waveforms of a saxophone and clarinet, respectively; reprinted with permission from UMAP module 588, "Music and the Circular Functions," by Dorothea Bone). A pure tone is represented by a sine wave. By combining a small number of pure tones in the proper proportions, a music synthesizer approximates the sounds of various instruments.

How are the proper proportions determined? The answer is: by using Fourier series!

The function $y = x$ generates one of the two basic non-pure waves built into synthesizers (the other, called a *square wave*, is discussed in the exercises). This

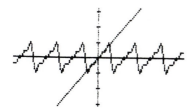

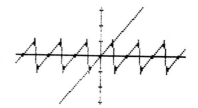

FIGURE 6.8a   FIGURE 6.8b

wave is called a *sawtooth wave*. The proportions of the various sine terms (called the *harmonics*) are crucial. In absolute value (with $2/\pi$ factored out) the size of the nth Fourier coefficient in the expansion of $f(x) = x$ is $1/n$. In the language of music synthesizers, the "harmonic content" varies as $1/n$ (recall that $\displaystyle\sum_{n=1}^{\infty} \frac{1}{n}$ is known as the harmonic series). By itself, the sawtooth wave represents an oboe-like sound, but it is easily modified (by varying the Fourier coefficients) to produce other familiar tones.

Exercises 6.3

In exercises 1-6, compute the Taylor polynomials of degrees 1, 2 and 3 of $f(x)$ centered at $x = a$ and sketch the graphs.

1. $\cos x$, $a = 0$

2. $x^4 - 1$, $a = 0$

3. $\sqrt{x + 1}$, $a = 0$

4. $\dfrac{1}{x+1}$, $a = 0$

5. $\sin x$, $a = \pi$

6. $\ln x$, $a = 1$

In exercises 7-10, compute the indicated Taylor polynomial and compare the graphs of $P_n(x)$ and $f(x)$. In each case, take $a = 0$.

7. $\cos x$, P_8

8. $\cos x$, P_{16}

9. $\dfrac{1}{x+1}$, P_9

10. $\dfrac{1}{x+1}$, P_{16}

In exercises 11-14, use Taylor series to argue that the limits are correct.

11. $\displaystyle\lim_{x \to 0} \frac{x - \sin x}{x^2} = 0$

12. $\displaystyle\lim_{x \to 0} \frac{x - \sin x}{x^3} = \frac{1}{6}$

13. $\displaystyle\lim_{x \to 0} \frac{e^x - 1 - x}{x^2} = \frac{1}{2}$

14. $\displaystyle\lim_{x \to 0} \frac{\ln x - (x - 1)}{(x - 1)^2} = -\frac{1}{2}$

In exercises 15-18, estimate $\int_0^1 f(x)\,dx$ using $P_3(x)$ for $\cos x$ (in exercises 15 and 17) or $\sqrt{x+1}$ (exercises 16 and 18).

15. $f(x) = \cos(x^2)$
16. $f(x) = \sqrt{x^2+1}$
17. $f(x) = \cos(x^3)$
18. $f(x) = \sqrt{x^3+1}$

In exercises 19-26, determine the first 4 terms of the Fourier series for $f(x)$ on $[-1,1]$ and graph the 4th partial sum. HINT: in exercises 19-22, use Examples 6-7.

19. $f(x) = x - 1$
20. $f(x) = x^2 - 1$
21. $f(x) = 2x - 1$

22. $f(x) = 3x^2 - 1$
23. $f(x) = |x|$
24. $f(x) = 3\sin(2\pi x)$

25. $f(x) = \begin{cases} -1 & x \le 0 \\ 1 & x > 0 \end{cases}$

26. $f(x) = \begin{cases} -1/2 & x \le -1/6 \\ \sin(\pi x) & -1/6 < x < 1/6 \\ 1/2 & x \ge 1/6 \end{cases}$

27. The Fourier series for the function in exercise 25 is the *square wave* which music synthesizers use. Describe the harmonic content of the square wave.

28. The Fourier series for the function in exercise 26 represents the *clipping* which a guitar amplifier does. The clipped function has nonzero harmonic content for all n, with a richer tone than a pure $\sin x$. Describe the harmonic content.

EXPLORATORY EXERCISE

Introduction

Fourier series is a part of the field of *Fourier analysis*, which is vital to many engineering applications. Fourier analysis includes Fourier transforms (you may have heard of the Fast Fourier Transform, or FFT) and various techniques for applying Fourier series to real world phenomena. We get an idea of how these techniques work below.

Problems

Given measurements of a signal (waveform), the goal is to construct the Fourier series of the signal function. We start with a simple version of the problem. Suppose the function has the form $f(x) = a_0/2 + a_1\cos\pi x + a_2\cos 2\pi x + b_1\sin\pi x + b_2\sin 2\pi x$ and we have the measurements $f(-1) = 0$, $f(-1/2) = 1$, $f(0) = 2$, $f(1/2) = 1$ and $f(1) = 0$. Plugging into f we get $f(-1) = a_0/2 - a_1 + a_2 = 0$, $f(-1/2) = a_0/2 - a_2 - b_1 = 1$, $f(0) = a_0/2 + a_1 + a_2 = 2$, $f(1/2) = a_0/2 - a_2 + b_1 = 1$

and $f(1) = a_0/2 - a_1 + a_2 = 0$. Note that b_2 never appears in an equation, and the $f(-1)$ and $f(1)$ equations are identical. We have 4 equations and 4 unknowns. Solve the equations [HINT: start by comparing the $f(1/2)$ and $f(-1/2)$ equations, then the $f(0)$ and $f(1)$ equations]. You should conclude that $f(x) = 1 + \cos \pi x$, and we have no information about b_2. To get b_2 we would need another function value. Thus, the number of measurements determines how many terms we can find in the Fourier series. Repeat the above for measurements $f(-1/2) = -1/2$, $f(0) = 0$, $f(1/2) = 1/2$ and $f(1) = 0$ and compare to the Fourier series in Example 7.

There is, fortunately, an easier way to determine the Fourier coefficients. Recall that $a_n = \int_{-1}^{1} f(x) \cos(n\pi x)\, dx$ and $b_n = \int_{-1}^{1} f(x) \sin(n\pi x)\, dx$. From the function values at $x = -1/2$, 0, $1/2$ and 1, we can estimate the integral. Which approximation rule gives the correct values of a_n and b_n in the above examples? Use this approximation rule to find the relevant constants given $f(-3/4) = -3/4$, $f(-1/2) = -1/2$, $f(-1/4) = -1/4$, $f(0) = 0$, $f(1/4) = 1/4$, $f(1/2) = 1/2$, $f(3/4) = 3/4$ and $f(1) = 0$. Again, compare to the series in Example 7.

Further Study

The general formulation of our work above is called the *inverse Fourier transform*. This can be found in numerous engineering mathematics books (see, for example, *Orthogonal Transforms for Digital Signal Processing* by Ahmed and Rao) although it is typically presented in terms of complex variables. For an enjoyable overview of several current applications of Fourier analysis, see *Visualization* by Friedhoff and Benzon, Abrams Publishers.

Bibliography

Adair, Robert K.: *The Physics of Baseball*, Harper and Row, New York, 1990.

Ahmed, N., and K.R. Rao: *Orthogonal Transforms for Digital Signal Processing*, Springer-Verlag, New York, 1975.

Barnsley, Michael: *Fractals Everywhere*, Academic Press, Boston, 1988.

Berg, Paul, and James McGregor: *Elementary Partial Differential Equations*, Holden-Day, San Francisco, 1966.

Bone, Dorothea: "Music and the Circular Functions," UMAP module #588, CO-MAP, Inc., Arlington, MA, 1983.

Boyce, William, and Richard DiPrima: *Elementary Differential Equations and Boundary Value Problems*, 5th edition, Wiley, New York, 1992.

Brancazio, Peter: *Sports Science*, Simon and Schuster, New York, 1984.

Brody, Howard: *Tennis Science for Tennis Players*, University of Pennsylvania Press, Philadelphia, 1987.

Buchanan, James and Peter Turner: *Numerical Methods and Analysis*, McGraw-Hill, New York, 1992.

Burden, Richard, and J. Douglas Faires: *Numerical Analysis*, 4th edition, PWS-Kent, Boston, 1989.

Churchill, Ruel, and James Brown: *Fourier Series and Boundary Value Problems*, 4th edition, McGraw-Hill, New York, 1978.

Conte, Samuel, and Carl deBoor: *Elementary Numerical Analysis*, 3rd edition, McGraw-Hill, New York, 1980.

Devaney, Robert: *An Introduction to Chaotic Dynamical Systems*, 2nd edition, Addison-Wesley, Redwood City, CA, 1989.

Feller, William: *An Introduction to Probability Theory and Its Applications*, volume 1, 3rd edition, John Wiley, New York, 1950.

Feller, William: *An Introduction to Probability Theory and Its Applications*, volume 2, John Wiley, New York, 1966.

Foerster, Mora and Amiot: "Doomsday, Friday, 13 November A.D. 2026," *Science*, volume 130, November 1960, pp. 1291-1295.

Friedhoff, R.M., and William Benzon: *Visualization*, Abrams, New York, 1989.

Gleick, James: *Chaos*, Penguin, New York, 1987.

Johnson, Lee, and R. Dean Riess: *Numerical Analysis*, 2nd edition, Addison-Wesley, Reading, MA, 1982.

Peitgen, H.-O., and D. Saupe, ed.: *The Science of Fractal Images*, Springer-Verlag, New York, 1988.

Raven, Francis: *The Mathematics of Engineering Systems*, McGraw-Hill, New York, 1966.

Thompson, J.M.T., and H.B. Stewart: *Nonlinear Dynamics and Chaos*, Wiley, New York, 1986.

Answers to Odd-Numbered Exercises

Section 1.1

1. 4.416666667   3. .3333333333   5. 12.64197531

7. 1.195228609   9. 24.2901664   11. 3.762195691

13. .984375 → 1.0   15. .8571428571 → 1.0

17. 2.976046176 → π 19. .5 → .5

21. $n = 39$ 23. $n = 32$ 25. 9°

33. P(1970) = 3.594 billion, P(1980) = 4.376 billion, P(1990) = 5.588 billion, and P(2035) = 4.131×10^{53} billion

Section 1.2

1. 3 roots   3. 1 root   5. no roots

7. 3 roots   9. no roots

21. simple oscillation, amplitude = $\sqrt{A^2 + B^2}$

23. vertical asymptotes: $x = -3$, $x = 1$; horizontal asymptote: $y = 0$

25. no vertical asymptotes; horizontal asymptote: $y = 2$

27. no vertical asymptotes; oblique asymptote: $y = x$

29. (.10365,1.4146), (16.4817,66.9268)   31. $\pm$ (.7391,.7391)

33. (.5671,.5667)   35. oblique asymptote: $y = 2x - 2$

37. no oblique asymptote, looks like a parabola for $|x|$ large.

39. Similar to $y = x^{n-m}$ (true for $n > m$).

41. It looks circular.

Section 1.3

1. ±1.414213562   3. −1, −2.831177207

5. $-1, 1.7099759467$

7. 1.409624004 , $-.6367326508$

9. $(x^2 - 2)(x^2 + 3)$

11. $10, \pm 1.414213562$

13. 0

15. max at $.183505$, min at 1.816496

17. min at 1.20507104

19. 14.1421-by-14.1421

21. same as exercise 19

23. $f(x)$ is nearly 0 for x between $-.2$ and $.2$.

25. $t = 1/3$; point of intersection: $(100,0)$

Section 2.1

1. $-1/2$ 3. does not exist   5. 0 7. $3/2$

9. 0 11. 0 13. $1/4; 1/\pi; 1/c$

15. does not exist, does not exist, 0

17. 0 19. 6 21. 1 23. 1

25. \$1100 (n=1); \$1102.50 (n=2); \$1103.81 (n=4); \$1104.71 (n=12);
\$1105.16 (n=365); \$1105.17 (continuous)

27. $.1, .08$

Section 2.2

1. $.25$; around $x = 50,001$

3. 1; around $x = 4 \times 10^8$

5. $1/6$; around $x = .0001$

11. 3; does not exist

13.

	$x = 1$	$x = 10$	$x = 100$	$x = 1000$
$\sin \pi x$	0	0	0	0
$\sin 3.14 x$	$.00159$	$-.0159$	$-.15859$	$-.99976$

Section 2.3

1. $\epsilon = .1 : \delta = .0333$; $\epsilon = .05 : \delta = .01666$ (Your values for δ could be smaller.)

3. $\epsilon = .1 : \delta = .316227766017$; $\epsilon = .05 : \delta = .2236$

5. $\epsilon = .1 : \delta = .39$; $\epsilon = .05 : \delta = .1975$

7. $\epsilon = .1 : \delta = .0465$; $\epsilon = .05 : \delta = .0241$

19. $.0066$

Section 3.1

1. 2 3. -1 5. 0 7. 1

9. 2 11. -1 13. 0 15. 1

19. 2 21. -1 23. 0

25. does not exist

Section 3.2

1. 0.0   3. 2.24   5. .136083

7. 0; 2.223244275   9. 1/2; .1111111111

11. 0; 2.325444263

13.

h	Forward	Backward	Centered
.1	.9950041653	.9950041653	.9950041653
.01	.9999500004	.9999500004	.9999500004
.001	.9999995	.9999995	.9999995

15.

h	Forward	Backward	Centered
.1	.0498756211	−.0498756211	0.0
.01	.004999875	−.004999875	0.0
.001	.0005	−.0005	0.0

17.

h	Forward	Backward	Centered
.1	.0990049834	−.0990049834	0.0
.01	.009999	−.009999	0.0
.001	.000999999	−.000999999	0.0

19.

h	Forward	Backward	Centered
.1	-.1484736111	.1484736111	0.0
.01	-.0149984585	.0149984585	0.0
.001	-.0014999985	.0014999985	0.0

21. All centered differences are 0.

23. $d/dx\,[\sin x^\circ] = d/dx\,[\sin(x\pi/180)] = (\pi/180)\cos(x\pi/180)$

25. .5   27. 1

29. 0   31. 1/2

Section 3.3

1.
$f(x)$	1	0	4	9
$T(x)$	−3	−1	3	5

3.
$f(x)$	$-1/\sqrt{2}$	$-1/2$	$1/2$	$1/\sqrt{2}$
$T(x)$	$-\pi/4$	$-\pi/6$	$\pi/6$	$\pi/4$

5.
$f(x)$	1.4142	1.7320	2	2.2360
$T(x)$	1.5	2.0	2.5	3.0

7.
$f(x)$	1.0772	1.2599	1.4812	1.7099
$T(x)$	1	1	1	1

9.
$f(x)$	.7071	.8660	.8660	.7071
$T(x)$	1	1	1	1

11. $f(x)$ .1353   .3678   2.7182   7.3890

   $T(x)$ -1 0   2   3

13. $-.41825$ 15. 2.05   17. 1.54326   19. $P - 2Px/R$

Section 3.4

5. 3.0   7. 2.0   9. 2.3346

11. 1.74339   13. 29.6923   15. 5

17. 2.51066   19. 1.7924   21. 3.4657 hr

23. 1, 1/2, 1/4, 1/8

25. « '(B-A)/H' EVAL 'N' STO 1 N FOR I EULER NEXT »

Section 4.1

1. -1.2184; BISECT-13 steps, NEWTON-4 steps, SECANT-5 steps

3. 2.0238; BISECT-13 steps, NEWTON-5 steps, SECANT-5 steps

5. $-.2975$; BISECT-13 steps, NEWTON-3 steps, SECANT-7 steps

7. -1.10485; BISECT-13 steps, NEWTON-3 steps, SECANT-7 steps

9. 1.73205   11. 2.1381   13. .73908

15. Division by 0 on second step; try $x_0 = .5$: $x = .4362$.

17. Division by 0 on first step; try $a = 1.5$ and $b = 2$: $x = 1.870495$.

19. Division by 0 with $x = 1$; multiply by $x^2 + 3x - 4$: $x = 1.414$.

21. $(x - 1)(x^2 + 4) = 0$ if $x = 1$; one Newton step.

23. $(x - 1)^2(x^2 - x + 3) = 0$ if $x = 1$; 14 Newton steps with TOL=.0001

25. $x = 1.847$, $d = 8 \times 1.847/5 - 2.153 = .8022$

27. N=12 (exercise 1), N=14 (exercise 3)

Exploratory - :0→I   :Lbl C   :A+I*(B−A)/100 → X   :X→S   :0→J   :Lbl D
:X−Y$_1$/Y$_2$ → X   :J+1→J   :If J<10   :Goto D   :Round(X,0)→X   :Disp X   :Disp
S   :Pause   :I+1→I   :If I<100   :Goto C

Section 4.2

1. 1 root   3. 2 roots   5. 2 roots

7. 4.917185925, 7.724153192, 11.08590173

9. -2.828427125, -2.0, 2.0, 2.828427125

11. -1.732050808, 1.732050808, 10.0

13. 1.0 (multiple), 2.0   15. 0.0 (multiple)

21. 2.25992105

Section 4.3

1. $f(1) = 2.0$ (absolute max), $f(-2) = -52.0$ (absolute min)
3. $f(1.732050808) = -25.57$ (absolute min),
   $f(-1.732050808) = 57.57$ (absolute max)
5. $f(-.4801899434) = -3.36$ (absolute min), $f(3) = 982$ (absolute max)
7. $f(0) = 3$ (absolute min), $f(2.236067978) = 3.16666$ (absolute max)
9. $f(1.264605496) = -.81844344$ (abs. max), $f(1) = f(3) = -1.0$ (abs. min)
11. $f(4) = -14.109$ (absolute min), $f(2.288929728) = 1.945$ (absolute max)
13. $f(0) = 0$ (absolute min), $f(2) = 4.27$ (absolute max)
15. $x = 3.9144$
17. $T(1.0739472) = 2.776$ hours (absolute min), saves .1069 hours (6.4 minutes)
21. .1877 sec   23. 116 ft/sec ($x = 105$)

Section 5.1

1. .39	3. 1/3	5. 3/4
7. 2.0	9. 1.0	11. 1.6
13. 2.6666	15. 3.464	17. 4.0
19. does not exist	21. exists	25. 1.166666

27. .5773502692, 1.527525232, 2.516611478

Section 5.2

1. 20 sqrt5/3   3. 11.68213261   5. 23.24835427
7. 2.875 (Trapezoid Rule); 2.86667 (Simpson's Rule)
9. 1.94375 (Trapezoid Rule); 1.9458333 (Simpson's Rule)

11.

N	Midpoint	Trapezoid	Simpson's
4	5.875	6.25	6.0
8	5.96875	6.0625	6.0
16	5.9921875	6.015625	6.0

13.

N	Midpoint	Trapezoid	Simpson's
4	.1896972656	.220703125	.2005208333
8	.1974029541	.2052001953	.2000325521
16	.1993494034	.2013015747	.2000020345

15. .58826   17. 1.4436   19. π (use RIEM)
23. All 3 are exact.
25. Only Simpson's Rule is exact.
27. $E_{2n} = S_{2n} + (S_{2n} - S_n)/15$

29. $\frac{1}{2} \ln |x| - \frac{1}{4} x^{-1} - \frac{1}{4} \ln(x^2 + 4) - \frac{1}{8} \tan^{-1}(x/2) + c$

31. $\frac{1}{2} \ln(x^2 + 1) + 2 \tan^{-1} x - \frac{1}{2} \ln(x^2 + 4) - \tan^{-1}(x/2) + c$

Section 5.3

1. 1.8458   3. 2.9205   5. .9545

7. .86466   9. $J = 2.133$, $v_f = 57$ mph   11. 2160

13. 67.848   15. 1.44829 (vs. 1.44797)   17. 2.96867π

19. 2.3364π 21. 670.886   23. -6400

25. .2   27. 7500   29. 9466.31

31. -1.973475, -5.608858, -8.358723; length of ground covered

Section 5.4

1. 1   3. 21.333333   5. π

7. $2\pi/3$ 9. 16π 19. .7854

21. π 23. .5435   25 $\pi/12$

27. 3.09748   29. .5708   31. n (n odd) or $2n$ (n even)

Section 6.1

1. 2.0   3. 0.0   5. 1.0, $N > 1/\epsilon$

7. 3.0, $N = 1$ 9. 2.0, $N > 1/\epsilon$ 11. 0.0

13. does not exist 15. 0.0   17. does not exist

19. 0.0   21. does not exist   23. factorials

Section 6.2

1. 8/3   3. diverges   5. $-3/5$

7. 1.202   9. -2.08 11. 2.578

13. $-.693$ 15. diverges   17. diverges

19. converges   21. diverges   23. converges

25. converges   29. 9

Section 6.3

1. $P_1(x) = 1; P_2(x) = P_3(x) = 1 - \frac{1}{2} x^2$

3. $P_1(x) = 1 + \frac{1}{2} x; P_2(x) = 1 + \frac{1}{2} x - \frac{1}{8} x^2; P_3(x) = 1 + \frac{1}{2} x - \frac{1}{8} x^2 + \frac{1}{16} x^3$

5. $P_1(x) = P_2(x) = -(x - \pi); P_3(x) = -(x - \pi) + \frac{1}{6}(x - \pi)^3$

7. $P_8(x) = 1 - x^2/2! + x^4/4! - x^6/6! + x^8/8!$

9. $P_9(x) = 1 - x + x^2 - x^3 + x^4 - x^5 + x^6 - x^7 + x^8 - x^9$

15. .9   17. .92857

19. $-1 + \frac{2}{\pi}[\sin \pi x - \frac{1}{2} \sin 2\pi x + \frac{1}{3} \sin 3\pi x - \frac{1}{4} \sin 4\pi x + ...]$

21. $-1 + \dfrac{4}{\pi}[\sin \pi x - \dfrac{1}{2} \sin 2\pi x + \dfrac{1}{3} \sin 3\pi x - \dfrac{1}{4} \sin 4\pi x + ...]$

23. $\dfrac{1}{2} + \dfrac{-4}{\pi^2}[\cos \pi x + \dfrac{1}{9} \cos 3\pi x + \dfrac{1}{25} \cos 5\pi x + \dfrac{1}{49} \cos 7\pi x + ...]$

25. $\dfrac{4}{\pi}[\sin \pi x + \dfrac{1}{3} \sin 3\pi x + \dfrac{1}{5} \sin 5\pi x + \dfrac{1}{7} \sin 7\pi x + ...]$

27. The harmonic content varies as $1/n$ for odd n.

Index